Regulating the Privately Rented Housing Sector

This book explores theory and practice in the complex policy area of privately rented housing in England, with a particular focus on environmental and public health. Bringing together a range of both academic and practicing experts in the field, it responds to the rapid growth and changing nature of the sector and considers the range of options available to local authorities in ensuring more effective regulation strategies.

This book:

- Creates a key, up-to-date professional resource for housing regulation based on road-tested academic course material.
- Breaks down strategies and practices to an implementational level.
- Provides impetus to leaders, practitioners, and students to both deliver and reflect on improved regulation.
- Explores responses to various stakeholder needs through the lens of protecting and supporting tenants.

This book will interest professionals working in public health, housing, and local authorities, as well as environmental health and housing academia. Students across environmental health, social work, nursing, and other disciplines will also find this appealing.

Jill Stewart has worked in housing for more than 30 years, as practitioner then academic in London Universities. She has published numerous books and papers and presented widely in the field, working with others on a range of funded projects. She co-created an innovative post-graduate CPD course with Russell Moffatt in advanced private-sector housing regulation to help develop the workforce. The second edition of her book *Environmental Health and Housing* was co-authored with Zena Lynch and published in 2018.

Russell Moffatt is a Chartered Environmental Health Practitioner specialising in private housing regulation and has worked on the front line in London for 20 years. He qualified with a B.Sc. (Hons) Environmental Health at Greenwich University in 2002 and achieved his Master's in Public Health at King's College, London, in 2006. Russell is a cofounder of Metastreet Ltd and co-created a post-graduate CPD course with Dr Jill Stewart in advanced private-sector housing regulation.

Routledge Focus on Environmental Health

Series Editor: Stephen Battersby, MBE, PhD, FCIEH, FRSPH

Tackling Health Inequalities
Reinventing the Role of Environmental Health
Surindar Dhesi

Statutory Nuisance and Residential Property
Environmental Health Problems in Housing
Stephen Battersby and John Pointing

Housing, Health and Well-Being
Stephen Battersby, Véronique Ezratty and David Ormandy

Selective Licensing
The Basis for a Collaborative Approach to
Addressing Health Inequalities
Paul Oatt

**Assessing Public Health Needs in a Lower
Middle Income Country**
Sarah Ruel-Bergeron, Jimi Patel, Riksum Kazi and Charlotte Burch

Fire Safety in Residential Property
A Practical Approach for Environmental Health
Richard Lord

COVID-19: The Global Environmental Health Experience
Chris Day

Regulating the Privately Rented Housing Sector
Evidence into Practice
Edited by Jill Stewart and Russell Moffatt

For more information about this series, please visit: https://www.routledge.
com/Routledge-Focus-on-Environmental-Health/book-series/CONENHE.

Regulating the Privately Rented Housing Sector

Evidence into Practice

**Edited by Jill Stewart
and Russell Moffatt**

LONDON AND NEW YORK

First published 2022
by Routledge
4 Park Square, Milton Park, Abingdon, Oxon OX14 4RN

and by Routledge
605 Third Avenue, New York, NY 10158

Routledge is an imprint of the Taylor & Francis Group, an informa business

© 2022 selection and editorial matter, Jill Stewart and Russell Moffatt; individual chapters, the contributors

The right of Jill Stewart and Russell Moffatt to be identified as the authors of the editorial material, and of the authors for their individual chapters, has been asserted in accordance with sections 77 and 78 of the Copyright, Designs and Patents Act 1988.

British Library Cataloguing-in-Publication Data
A catalogue record for this book is available from the British Library

Library of Congress Cataloging-in-Publication Data
A catalog record for this book has been requested

ISBN: 978-1-032-15969-0 (hbk)
ISBN: 978-1-032-15971-3 (pbk)
ISBN: 978-1-003-24653-4 (ebk)

DOI: 10.1201/9781003246534

Typeset in Baskerville
by Apex CoVantage, LLC

This book is dedicated to all those who lost their lives to Coronavirus, including the much-loved Stanley Field.

Contents

Contributors   ix
Acknowledgements   xii

Introduction   1

1 **Introduction to understanding the private rented sector**   4
  JULIE RUGG

2 **The private rented sector and the meaning of home**   14
  JILL STEWART AND ZENA LYNCH

3 **Perspectives on the regulatory framework and
  intervention**   22
  JILL STEWART AND ZENA LYNCH

4 **Partnerships of prevention: beyond regulation**   30
  ELLIS TURNER

5 **Practical problems before the courts and tribunals**   40
  DAVID SMITH

6 **The shadow private rented sector examined**   48
  BEN REEVE-LEWIS

7 **Making more effective intervention choices**   58
  DAVID BEACH

8 **The Housing Health and Safety Rating System (HHSRS): a practitioner's perspective**   66
ALEX DONALD

9 **Regulating houses in multiple occupation (HMOs)**   76
LOUISE HARFORD AND KEVIN THOMPSON

10 **Housing Act 2004 property licensing schemes**   88
HENRY DAWSON AND RICHARD TACAGNI

11 **Advanced regulatory skills and practical evidence gathering**   97
PAUL OATT

12 **Embedding research on public health and housing into practice**   106
MATT EGAN, CHIARA RINALDI, JAKOB PETERSEN, MAUREEN SEGUIN, AND DALYA MARKS

13 **Developing effective PRS regulatory strategies**   116
RUSSELL MOFFATT

   Conclusions   123

   Index   125

Contributors

Alex Donald is involved in prosecuting rogue landlords and agents at the London Borough of Barking and Dagenham. Having trained in HHSRS and advanced PRS regulation, he has also been elected into public office as a Councillor at the London Borough of Havering.

Ben Reeve-Lewis has been a tenancy relations officer since 1990, prosecuting harassment and illegal eviction and addressing rogue landlord and criminal activity in the PRS. In 2015 he co-founded Safer Renting, an independent, non-profit tenants' rights advocacy service.

David Beach is housing manager at London Borough of Waltham Forest, recognised for pioneering enforcement approaches. He has published some of his work and is a visiting lecturer on the Middlesex University advanced private-sector housing regulation course.

David Smith obtained his doctorate in international relations before retraining as a solicitor and specialising in residential landlord and tenant law and the regulation of rental property. He has extensive experience working with landlords, agents, devolved governments, and LAs and has written several books on different aspects of residential property law.

Ellis Turner is a senior lecturer at the University of West England, having previously worked as an EHP in London for around 20 years. Ellis's research interests include hoarding behaviour and evaluating interventions. Ellis is a trustee of the Association of London Environmental Health Managers and Hoarding UK.

Henry Dawson is a Chartered EHP and Senior Lecturer at Cardiff Metropolitan University in housing and health. He has more than a decade of experience in LA housing enforcement. His research interests are around licensing and regulation of PRS properties.

Jill Stewart has worked in housing for more than 30 years as practitioner then academic and has published and presented widely. She co-created and leads the post-graduate CPD course with Russell Moffatt in advanced private-sector housing regulation to help develop the PRS workforce.

Julie Rugg is a senior research fellow at the Centre for Housing Policy, University of York. She has undertaken multiple research projects with a focus on the PRS, including two major reviews: The Private Rented Sector: Its Contribution and Potential (2008) and The Evolving Private Rented Sector: Its Contribution and Potential (2018). She is interested in landlord and tenant behaviours and the impact of formal and informal regulatory frameworks.

Kevin Thompson has worked in PRS regulatory services for more than 30 years for a number of London boroughs. He was with the CIEH policy team with a housing brief and with LACORS, where he co-ordinated national guidance on fire safety standards for housing. Kevin has undertaken consultancy work on PRS strategy and business improvement. He is currently head of private-sector housing at Hackney Council.

Louise Harford is an EHP and has worked in PRS for more than 13 years. She specialises in HMO regulation and enforcement, including property licensing. Louise has a special interest in joining up services for better outcomes.

Matt Egan, **Chiara Rinaldi**, **Jakob Petersen**, **Maureen Seguin**, and **Dalya Marks** are researchers from London School of Hygiene & Tropical Medicine (LSHTM) who work together on place-centred public health projects. This work includes research into PRS and selective licensing. Egan is a professor of public health; Rinaldi is starting a PhD fellowship; Petersen and Seguin are research fellows; and Marks is an associate professor and also works as a public health strategist in Camden and Islington. These authors are funded by the National Institute for Health Research (NIHR) School for Public Health Research (SPHR), Grant Reference Number PD-SPH-2015. The views expressed are those of the authors and not necessarily those of the NIHR or the Department of Health and Social Care.

Paul Oatt is a Chartered EHP with 20 years' experience in LA regulation to management level. He has an MSc in public health from the LSHTM. Paul has taught at Cardiff Metropolitan University and Middlesex University and authored Selective Licensing: The basis for a collaborative approach to addressing health inequalities (2020).

Richard Tacagni is a Chartered EHP with more than 25 years' housing experience. He is managing director of London Property Licensing, an independent housing consultancy that provides advice to the lettings industry and LAs on PRS regulation. He acts as an expert witness in housing regulation, certified by Cardiff University School of Law and Politics (2020).

Russell Moffatt is a Chartered EHP specialising in PRS regulation and has worked on the front line in London for 20 years. He achieved his Masters in Public Health at King's College, London, in 2006. Russell is a cofounder of Metastreet Ltd and co-created the post-graduate CPD module in advanced private-sector housing regulation.

Zena Lynch is an associate professor at the University of Birmingham and is M.Sc. Environmental Health programme leader. Prior to this she was health policy lead for the West Midlands Regional Assembly. Zena worked for many years as an EHP, latterly focussing in the housing and public health arena.

Acknowledgements

We would like to thank Dr Rob Couch for permission to reproduce his diagram: The governance model of regulation by local government EHPs.

Introduction

The private rented housing sector (PRS) has grown substantially in recent years, housing some 1.7 million households, including children (Baxter and Murphy, 2019). Landlords, and indeed agents, are a diverse group, and whilst some provide good accommodation, many do not. The English Housing Survey reports that it contains the poorest conditions overall, with 13% having at least one Category 1 hazard, although this is reported as decreasing (MHCLG, 2019, 2020). The PRS can be expensive and insecure, creating and exacerbating inequalities and inequities. There are financial and other costs to the public purse in benefits as well as in wider health and social costs.

The PRS is not always the tenure of choice, and the issues are far wider than just supply and demand. Many tenants feel trapped across the life-course, from young families to those ageing in place. The problem of housing churn does nothing to help community stability, let alone provide a basic platform for anyone to be able to thrive and maximise their life opportunities. Houses in multiple occupation (HMOs) are particularly challenging, and there is little research and evidence around effective strategies and interventions in this form of accommodation. There is also the question of what home means or can be for many PRS tenants.

We know much about the relationship of poor housing conditions to health and safety and, indeed, quality of life. What we are less clear about is how we can effectively intervene. Local strategies should address both individual properties and the local environment and community, focus on priority health needs, identify the particular challenges, and apply evidence-based responses. The living environment should protect and improve our health, and we should not have to adapt to suit poor living accommodation and environments. The socio-environmental effects of neighbouring properties and the wider community can be acute; for example, anti-social behaviour and disorder are sometimes linked with high concentrations of poorly managed PRS. Area-based interventions, including property licensing, can help.

DOI: 10.1201/9781003246534-1

However, there is a lack of research and evidence around regulating the PRS effectively and helping ensure that tenants are adequately protected. We are also mindful, as editors of this book, that the PRS is not always appropriate and would like to acknowledge the need for a greater supply of social housing. PRS interventions can be complex and take considerable time, leaving tenants in precarious and stressful situations when they need a rapid remedy. We can help mitigate some of the worst effects by working in effective partnerships to address wider factors such as poverty and deprivation. Housing should make people's lives better, not worse.

There are multiple laws and regulations to help us in what we must do to address conditions and management but also the wider living environment. The Housing Act 2004 provides duties and powers around the Housing Health and Rating System (HHSRS) (at the time of writing under review) as well as mandatory and discretionary licencing and other management options. The Housing and Planning Act 2016 supports this with additional powers for Banning Orders, Rogue Landlord Database, Civil Penalties, and an extension for Rent Repayment Order provisions. There are also regulations which apply to England on a range of health and safety requirements.

We need to better understand our local housing stock and optimise its use and potential. We need to continue to develop evidence-based strategies to deliver better public health outcomes. We need to make better use of legislation and have a suitably qualified and diverse workforce. We need to embrace technology and develop the role data science can play. We need to become increasingly proactive and use evidence to support what we deliver at the front line of practice and at the local and national policy levels.

Against this backdrop there has been a plethora of publications in recent years focusing on the privately rented sector, costs associated with poor conditions, and workforce challenges. This comes not just from think tanks and universities but also from House of Commons Briefings, recognising the multiple challenges faced (for example, Barton and Kenny, 2018; Rugg, 2020).

This book explores how we can provide more comprehensive, longer-term approaches in regulating the PRS. We use the term "local authority (LA)" widely, to include local housing authority as appropriate. We have generally used the term Environmental Health Practitioner (EHP) drawing from wider literature, but for the purposes of this book, this term also includes others working in private-sector housing regulation teams carrying out equivalent housing work. Drawing from experts in the field and what we have learnt from delivering the advanced private-sector housing regulation module at Middlesex University, the synopsis of our book is as follows:

Chapter 1 – Introduction to understanding the private rented sector
Chapter 2 – The private rented sector and the meaning of home

Chapter 3 – Perspectives on the regulatory framework and intervention
Chapter 4 – Partnerships of prevention: beyond regulation
Chapter 5 – Practical problems before the courts and tribunals
Chapter 6 – The shadow private rented sector examined
Chapter 7 – Making more effective intervention choices
Chapter 8 – The Housing Health and Safety Rating System: a practitioner's perspective
Chapter 9 – Regulating houses in multiple occupation (HMOs)
Chapter 10 – Housing Act 2004 property licencing schemes
Chapter 11 – Advanced regulatory skills and practical evidence gathering
Chapter 12 – Embedding research on public health into practice
Chapter 13 – Developing effective private-sector regulatory strategies
Conclusions will then be drawn.

References

Barton, C. and Kenny, C. (2018). *Health in private-rented housing.* Number 573 April 2018, Houses of Parliament, Parliamentary Office of Science and Technology (POST) Research Briefing.

Baxter, D. and Murphy, L. (2019). *Sign on the dotted line? A new rental contract final report.* London: Institute of Public Policy and Research.

MHCLG. (2019). *English private landlord survey main report.* London: MHCLG.

MCHLG. (2020). *English housing survey headline report Dec 2019–2020.* London: MHCLG.

Rugg, J. (2020). *London Boroughs' management of the private rented sector.* London: Trust for London.

1 Introduction to understanding the private rented sector

Julie Rugg

Introduction

The private rented sector (PRS) is a highly complex part of the housing market, containing numerous niche markets and a range of suppliers and demand groups, leading to considerable diversity at a local level. At least five major government departments frame major regulations for the PRS, but there is limited co-ordination on policy objectives.

According to the Survey of English Housing, the PRS accommodated 4.4 million households in 2019/20. Strong growth in the years immediately following the global financial crisis has plateaued, and over the last five years the sector has contracted slightly (see Table 1.1). Overall, it appears that the number of newly created households seeking accommodation in the PRS is falling.[1] In part, this contraction reflects a reduction of numbers in some demand groups including economic migrants and growing numbers of households accessing home ownership.

Table 1.1 Trends in tenure: households

	Households (,000s)			Households (%)		
	Owner occupation	Social renting	Private renting	Owner occupation	Social renting	Private renting
2015/16	14,330	3,918	4,528	62.9	17.2	19.9
2016/17	14,444	3,947	4,692	62.6	17.1	20.3
2017/18	14,784	3,958	4,530	63.5	17.0	19.5
2018/19	15,018	3,963	4,552	63.8	16.8	19.3
2019/20	15,362	3,978	4,438	64.6	16.7	18.7
	+1,032	+60	−90			

Source: English Housing Survey, Table FT1101 (S101): Trends in tenure.

DOI: 10.1201/9781003246534-2

Supply of privately rented property

In England, landlordism is often likened to a 'cottage industry'. The nature of the housing market and its regulatory framework means that it is relatively easy to purchase property and let it to someone else. Many landlords are individuals with small portfolios, where letting property is not a principal source of income. A regularly updated English Private Landlord Survey offers information on the characteristics of landlords and letting agents, their portfolios, source of finance and letting preferences (MHCLG, 2018). Recent qualitative research indicates that smaller landlords can be categorised using a four-part classification:

- *Accidental or incidental landlords* let inherited or 'spare' property following relationship formation or temporary movement to take up a work opportunity. Property ownership includes just one or two properties. Letting might take place in the short or medium term until decisions are made about sale.
- *Investment landlords* are in paid employment but also purchase property to let to augment income or as a means of saving money, often for use in retirement or to help children onto the property ladder. Investment landlords often draw on their own housing equity and make use of buy-to-let mortgage finance. Letting property is regarded as a long-term activity for this group.
- *Portfolio landlords*, whose sole income comes from letting property, will also be using buy-to-let finance and make use of commercial loans. Some may have inherited a portfolio, and multi-generation letting is not uncommon: it can be a 'family business'. There is a strong association between portfolio landlordism and a background in property trades.
- *Business landlords* are high-net-worth individuals, where letting property is one of a number of business interests and who directly employ other individuals to manage the lettings.

(Rugg and Wallace, 2021)

Over time, the proportion of landlords with larger holdings has increased: the Private Landlord Survey 2018 found that 16.7 per cent of landlords had five or more properties; in 2010, this proportion was 4.9 per cent.[2] However, it remains the case that 52 per cent of households are letting from a landlord with four or fewer properties, and 48 per cent are letting from a landlord with five or more properties (MHCLG, 2019: 10).

Other organisations and institutions also let property in the private rental market. Housing associations and local authorities are increasingly looking to let property at market rates and using assured shorthold tenancies to secure rental profits to support their social objectives. Crook and Kemp (2019) calculated that the largest housing associations were increasing their holdings of property let at market rents; by 2015, ten housing associations had 500+ market rental properties (Crook and Kemp, 2019). Many local authorities have set up housing companies which in part rely on the letting of property at market rates (Morphet and Clifford, 2019), although the scale of this contribution to PRS stock is unclear.

In addition, the private rented sector also includes a 'build to rent' (BTR) element. Industry definition of BTR indicates a number of core components:

> Typically, it involves an institution, such as a pension fund, investing in providing private rented sector homes. Unlike with traditional, individually let private rented homes, Build to Rent developments are managed as a whole, providing communal facilities and social activities to everyone living in the development, through technology platforms and/or staff available on site.[3]

The intention is for the property to remain in the rented sector, although ownership might change hands on the global property market. The British Property Federation indicated that in June 2018 20,863 units had been completed (Rugg and Rhodes, 2018: 39); by Q2 2021 this figure had increased to more than 62,274.[4] BTR attracts substantial overseas investment. Initially, funding focussed on higher-density city-centre dwellings aimed at 'young professionals', but in recent years the market has started to expand into the creation of suburban, rent-only estates for families.[5]

Landlords might make use of letting agents to arrange a letting or to take over management of a property in its entirety. The Private Landlord Survey 2018 indicated that just under half of surveyed landlords used a letting agent, although just one in ten used an agent for both letting and management services. Landlords were most likely to use an agent to arrange letting the property and then manage the tenancy themselves (MHCLG, 2019: 20ff). As an element of the PRS, letting agents are highly diverse and include large-scale national chains, franchises, smaller independent regional or local businesses and businesses operating entirely online. Some landlords also informally manage property on behalf of friends or relatives. Recent legislation has increased the level of regulation aimed at letting agents: from 2014, letting agents were required to sign up to one of two designated redress schemes; and the Tenant Fees Act 2019 limited the fees that can be charged to tenants.

Demand groups

The PRS has traditionally served multiple housing needs in a more readily accessible fashion than either home ownership or social housing. It is often useful to view housing consumption as taking place dynamically across the life course (Beer and Faulkner, 2011). Demand for more flexible tenures often happens during transitions: the move out of the parental home, household formation and periods of temporary work or study. Key demand groups for rental property therefore include students; younger couples making their first home together; households transitioning from one owner-occupied property to another; and economic migrants (Rugg and Rhodes, 2018; Perry, 2012). Ten per cent of all PRS households comprise lone individuals sharing with other lone individuals: in owner occupation and social housing, this household type is less than 2 per cent.[6] Houses in multiple occupation (HMOs) are almost always privately rented.

However, the private rented sector is now often meeting demand from groups that might previously have been able to secure property in social housing, including families with young children and households on lower incomes. It is increasingly the case that couples are starting their families whilst living in the PRS. Analysis of the Family Resources Survey indicates a marked growth since 2008/9 in the proportion of privately renting couples with dependent children and a particular increase in the proportion of households containing children under the age of five (Rhodes and Rugg, 2018). Indeed, households containing pre-primary school – age children are more likely to be living in the PRS than any other tenure. This trend underlines the importance of property condition in the PRS given long-term likely impact on children's health.

Table 1.3 Private renters: proportion of income spent on rent, 2019/20, by region

	Mean	*Median*
North East	**33.7**	**26.5**
North West	**29.0**	**25.0**
Yorkshire and the Humber	**24.4**	**22.4**
East Midlands	**24.8**	**20.9**
West Midlands	**26.8**	**22.0**
East	**28.7**	**26.2**
London	**42.2**	**32.2**
South East	**32.5**	**26.7**
South West	**31.4**	**28.3**

Source: English Housing Survey: Private Rented Sector, Annex Table 2.7

The PRS is now also more likely to be accommodating households on lower incomes who are seeking long-term tenure at an affordable rent. In many – but not all – cases these households will be receiving benefit to help pay some or all their rent. This group is particularly vulnerable in terms of household composition, age and financial resilience. For example, in the PRS overall, around a tenth of households comprise lone parents; where households receive housing benefit, this figure is just under a third. Similarly, single households over pension age comprise around 5 per cent of all PRS households but closer to 10 per cent again amongst housing benefit recipients (Rugg and Wallace, 2021). Private renters are more likely to be in financial difficulties: according to the Family Resources Survey 2017/18, just over a fifth were unable to keep up with their bills.

Demand for rental property takes place in the open market, but it is also the case that a 'mediated market' exists, where properties for rental are procured by statutory and third-sector agencies to meet the needs of a range of target groups who are marginalised in that market. The Homelessness Reduction Act 2017 expanded local authority responsibilities to prevent and tackle homelessness by removing restrictions relating to intentionality and whether the household was deemed to be in priority need. The limited availability of property in the social rented sector has increased local authority reliance on the PRS. In Q1 2021, just under a third of homelessness relief was met through placing people in privately rented accommodation.[7] Local authority reliance on the PRS for temporary accommodation (TA) is also heavy: in the same quarter, an estimated 59 per cent of TA need was met through property procured from private landlords.[8] The Probation Service and local authority social care services also seek to lease properties from the PRS, as does a wide range of housing-related charities. It is not necessarily the case that a third-party intermediary is in a position to ensure that property is free from hazards, particularly in areas of limited housing supply and acute demand from homeless households.

Geography and rents

There is regional differentiation in terms of the size of the PRS and its trends in terms of growth, stagnation or decline. There can be substantial variation, year on year, in the number of privately renting households in a given region. Table 1.2 indicates difference in the number of households across the nine English regions between the period 2014/15 and 2019/20. This time period saw a 5.6 per cent increase in households in all tenures. West Midlands and London saw a large increase in numbers above that percentage, but the North East, East and South West all contracted. These

Table 1.2 All PRS households (n., 000s) across regions: change between 2014/15 and 2019/20

	2014/15	*2019/20*	*2014/15–2019/20*
North East	185	177	−4.3
North West	523	555	6.1
Yorkshire & the Humber	430	471	9.5
East Midlands	339	362	6.7
West Midlands	363	408	12.4
East	428	423	−1.2
London	898	994	10.7
South East	691	614	−11.1
South West	421	435	3.3

Source: English Housing Survey, Annex Table 1.2: Tenure, by region, 2003–04 to 2019–20

figures indicate that any national 'story' of the PRS is unlikely to hold true for all parts of the country.

At the more localised level, the PRS is configured according to patterns of supply and demand where dominant demand groups may frame common practices in the market. For example, a smaller city with a large higher education institution may well have a strongly defined student housing market. In many cities with a large student population, private-sector halls of residence will have added a substantial number of privately rented bedspaces to the locality. This provision often displaces demand from 'street' properties, provoking landlords to switch their target group or leave the market altogether. In some locations the proportion of households receiving benefit help to pay the rent can be a dominant feature, with landlords adapting their management practices accordingly (Rugg and Wallace, 2021).

Other factors shape the nature of the local PRS including the nature of local residential property. Coastal towns often have a PRS dominated by HMOs or smaller studio apartments carved out of large Victorian or Edwardian 'villa' properties. The PRS in rural locations is more likely to be concentrated in the larger market towns, although the incidence of economic migration linked to the agricultural labour market will create time-limited and spatially concentrated pockets of extremely heavy demand. Here, overcrowding may well be in evidence alongside the use of non-standard structures for residential purposes.

Private renting encompasses lettings at the very top of the market, in super-prime rental properties starting at around £5,000 a week (Hall, 2021) and informal lettings to family members at peppercorn rents. Table 1.3 indicates English Housing Survey data on the proportion of income spent on rent, by region, using the median measure. Again the table shows substantial

Table 1.3 Private renters: proportion of income spent on rent, 2019/20, by region

	Mean	*Median*
North East	33.7	26.5
North West	29.0	25.0
Yorkshire and the Humber	24.4	22.4
East Midlands	24.8	20.9
West Midlands	26.8	22.0
East	28.7	26.2
London	42.2	32.2
South East	32.5	26.7
South West	31.4	28.3

Source: English Housing Survey: Private Rented Sector, Annex Table 2.7

regional difference with regard to affordability: in the East Midlands region, private renters paid around a fifth of their income on rents; in London, this proportion was closer to one third.

Tenancy length

More than 80 per cent of lettings in the PRS are assured shorthold tenancies (AST). This type of tenancy was introduced by the Housing Act 1996 and has gradually superseded all other tenancy types: regulated tenancies now form less than one per cent of all tenancies. In 2019/20, 54 per cent of ASTs had an initial letting period of twelve months and a further 31 per cent were let, initially, for six months.[9] After the initial letting period, the tenancy can continue to become a periodic tenancy if no further term is fixed. The Housing Act 1996 defines mandatory and discretionary grounds for possession where a landlord and tenant are in dispute over a tenancy coming to an end. At time of writing, the government had made a commitment to the abolition of Section 21 of the Housing Act 1988, which permits landlords to terminate a tenancy after its initial fixed term without having to state a reason.

Tenancy length has become more extended as a higher proportion of tenants seek a longer-term home in the sector. In 2019/20, 37 per cent of tenants were in a tenancy that had lasted two years or less, but 51 per cent had lived in the same property for between two and ten years. Eleven per cent of renters had lived in the same property for ten years or more.[10]

Regulating the private rented sector

The PRS is subject to a wide range of regulatory interventions generated by government departments including the Department for Levelling Up, Housing and Communities (DLUHC);[11] the Department for Work and Pensions (DWP); the Department for Business, Energy and Industrial Strategy (BEIS); the Ministry of Justice (MoJ); and HM Revenue and Customs (HMRC). These regulatory interventions do not necessarily combine to demonstrate coherent objectives or strategy for the PRS. Over the last twenty years, successive governments have favoured piecemeal approaches to tackling PRS-related problems relating to property condition and management (Rugg and Rhodes, 2018).

There are perhaps three major interconnecting areas of regulation which together have the most substantial impact on the sector. The DLUHC oversees the operation of the Housing Act 2004 and the Housing and Planning Act 2016. These regulations frame local authority enforcement activity, although recent research has indicated that there is considerable variation in the tenor of that activity (Harris et al., 2020). Austerity measures have reduced the number of enforcement officers in many local authorities.

DLUHC also has responsibility for homelessness and homelessness prevention. The Homelessness Act 2002 requires each local authority to produce a regularly updated homelessness strategy (LGA, 2019). This strategy should include a review of measures to tackle and prevent homelessness. The PRS is central to both the causes and amelioration of homelessness. The ending of an assured shorthold tenancy is a major cause of homelessness, largely because lower-income tenants face problems in securing alternative affordable accommodation in a timely fashion. At the same time, local authorities are making increasing use of the PRS to discharge their homelessness duties. Housing officers should work closely with environmental health practitioners to ensure that households are helped into good-quality properties, but this is not always the case (Rugg, 2020).

A third area of regulation sits within the purview of the DWP and relates to welfare assistance with rental payments. The Welfare Reform Act 2012 introduced Universal Credit, which replaced a system of local-authority administered housing assistance with a centrally controlled framework accessed by applicants on line or via Jobcentre Plus. Applicants are paid the local housing allowance (LHA) permitted for their household size, irrespective of the rent charged by the landlord. LHA rates are set within the boundaries of broad rental market areas and in theory should reflect the thirtieth percentile rent. However, LHA rates have been subject to caps and freezes irrespective of change in localised rents and which has created a dislocation between market rents and LHA rates.

These three sets of regulations have tended to operate together to worsen the problems relating to letting at the lower end of the PRS: UC has reduced the supply of more affordable property as landlords exit the housing benefit market; homelessness measures have increased reliance on the PRS sometimes irrespective of property condition; and local authorities are not necessarily consistent in their approaches to enforcement, leaving tenants subject to a 'postcode lottery'.

Conclusion

This introductory chapter has set out some of the principal characteristics of the PRS and its regulatory framework. Further chapters in this text will explore and examine some of the points raised here. The sector is currently going through a period of flux in terms of changes in both the characteristics of property supply and demand. In the absence of political will to create a more coherent objective for the sector, the government is likely to continue bringing piecemeal interventions to the market, adding to the complexity of an already densely convoluted regulatory landscape.

Notes

1 See www.gov.uk/government/statistics/english-housing-survey-2019-to-2020-headline-report, Section 1 Household Tables, Figure 1.10: Household moves by tenure, 2019–20.
2 www.gov.uk/government/statistics/private-landlords-survey-2010, AT2.2a; www.gov.uk/government/publications/english-private-landlord-survey-2018-main-report, AT1.4. Accessed 8 Sep 2021.
3 https://buildtorent.info/consumers/, accessed 8 Sep 2021.
4 https://bpf.org.uk/about-real-estate/build-to-rent/, accessed 8 Sep 2021.
5 https://btrnews.co.uk/build-to-rent-moves-to-the-burbs/c, accessed 20 Sep 2021.
6 See www.gov.uk/government/statistics/english-housing-survey-2019-to-2020-headline-report, Section 1 Household Tables, Annex Table 1.3: Demographic and economic characteristics.
7 See www.gov.uk/government/statistical-data-sets/live-tables-on-homelessness, Table R2: Accommodation secured at relief duty end.
8 See www.gov.uk/government/statistical-data-sets/live-tables-on-homelessness, Table TA1: Households in temporary accommodation at end of quarter.
9 English Housing Survey Private Rented Sector Report, Annex Table 1.16: Tenancy types in the private rented sector, 2019–20.
10 English Housing Survey Private Rented Sector Report, Annex Table 3.1: Length of residence in current accommodation, 2019–20.
11 Until September 2021, the Ministry of Housing, Communities and Local Government.

References

Beer, A. and Faulkner, D. (2011). *Housing transitions through the life course: Aspirations, needs and policy*. Bristol: Policy Press.
Crook, A.D.H. and Kemp, P. (2019). In search of profit: Housing association investment in private rental housing. *Housing Studies*, 34, pp. 666–687.

Hall, Z.D. (2021). Mansion for rent: Will London's super-rich pay up to £100,000 a week? *Financial Times*, 4 August 2021. www.ft.com/content/c8fee088-9aa7-4d48-bb81-e0e2cb87939d (accessed 21 September 2021).

Harris, J., Cowan, D., and Marsh A. (2020). *Improving compliance with private rented sector regulation*. Bristol: UK Collaborative Centre for Housing Evidence.

LGA. (2019). *Making homelessness strategy happen: Ensuring accountability and deliverability*. London: LGA.

MHCLG. (2019). *English private landlord survey 2018 main report*. London: MHCLG.

Morphet, J. and Clifford, B. (2019). *Local authority direct delivery of housing: Continuation research*. London: UCL.

Perry, J. (February 2012). UK migrants and the private rented sector. *JRF Findings*. www.metropolitan.org.uk/images/Uk-Migrants-and-the-Private-Sector-Findings.pdf (accessed 21 September 2021).

Rhodes, D. and Rugg, J. (2018). *Vulnerability amongst low-income, households in the private rented sector in England*. York: University of York, Centre for Housing Policy and Nationwide Foundation.

Rugg, J. (2020). *London Boroughs' management of the private rented sector: A briefing paper*. London: Trust for London.

Rugg, J. and Rhodes, D. (2018). *The evolving private rented sector: Its contribution and potential*. York: Centre for Housing Policy.

Rugg, J. and Wallace, A. (2021). *The supply of property to the lower end of the private rented sector*. York: Centre for Housing Policy.

2 The private rented sector and the meaning of home

Jill Stewart and Zena Lynch

Introduction

This chapter considers literature relating to the context and meaning of home for private renters, asking what it should be and the role EHPs might play in addressing the complex issues faced.

What is 'home' for private renters?

Home is something we all have a notion of, yet it can be tricky to define. It is a base where we feel safe, do routine things and have access to the wider environment. It is where we learn 'the rules', behaviours, have family and friends round. 'Home' for some can also of course be the opposite, somewhere that is insecure, imprisoning, frightening, in poor condition and badly managed. A home should provide foundations for living and be adaptable to changing needs across the life course.

We spend much of our time at home in relation to other environments, so it is particularly important that we feel secure and comfortable. The 'ontological security' it provides, however, is more commonly linked to home ownership rather than privately renting (Hiscock et al., 2010; Saunders, 2007). Home is part of emotional and material identity, social relations, gender issues, class and meaning. There are mundane things about home: the day-to-day lived reality of normal, routines, the neighbours and getting to other places (school, shops, doctor, social care, leisure facilities) with relative ease.

Home is about attachment. Many try to personalise their home (where they are able to) and from this access facilities in the wider environment that extend their sense of place, community and belonging or − conversely − anomie. The home is a resource providing the foundations of resilience that can help or hinder our lives. It should be a positive resource that adds quality and value, but this is not always the case, with some things outside of the tenant's control, such as length of stay, making it difficult to feel at home. These

DOI: 10.1201/9781003246534-3

circumstances require personal and psychological adjustment and can be stressful. Home means different things to different people, with sometimes opposite connotations.

In 1986 Sixsmith argued that the concept of 'home' (then) lacked a theoretical framework but found that different types of home exist and that different meanings for home co-exist. Somerville (1997) proposes that such a theory needs to explain the meaning of home as a complex, multi-levelled and multi-dimensional construct in the context of wider relations and settings. Home is physical, psychological and socially constructed in real and ideal forms. There are multiple meanings around the home, each of which needs to be understood as physical, psychological and social constructs related to wider factors.

More recently Hulse and Haffner's (2014) work in Australia tells us that whilst research around owner occupation presents the meaning of home as multidimensional, the PRS and home relationship remains under-theorised and -researched, with challenges presented in a single dimensional manner, such as property rights (e.g. security of tenure) or public policy issues (e.g. rent control), with research gaps. It can be very difficult for some private renters to have good housing and wellbeing outcomes in their homes because of a general lack of stability (Smith et al., 2014).

Home is also about one's identity and autonomy, emotional security, feelings of control over the housing. Home is also about the suitability or adaptability of the housing environment to be able to respond to changing needs, caused by having a family, ageing or disability. As Easthope (2004) argues, home is a particular type of 'place' which correlates to identity and psychological well-being and has an important role to play in housing research but is often neglected. Clapham (2010) describes subjective feelings around home: happiness, self-esteem and positive identity for residents. He asserts that policy currently sees housing as units of accommodation rather than homes and argues that housing policy should favour happiness and well-being as primary objectives. Barratt and Green (2017) focus on the relationship between place of residence and a person's identity, linking to the creation of 'self' and the significance of this in terms of a person's self-perception of well-being, which is important in terms of life course outcomes.

Home and security

Much of the literature is around positive meanings of home in its relation to security, whether for owner occupiers or tenants benefitting from secure tenancy agreements. Security affects rights, sense of control and well-being (see, for example, Easthope, 2014). There is little literature around English tenancies and the relationship to a sense of home, which can create a power imbalance in the landlord/tenant relationship. Tenants may not risk requesting repairs for fear of eviction or rental increase, and landlords may not know the standards that are required (Barton and Kenny, 2018). Overall the tenant's primary security and sense of control over their home is affected, leaving a constant state of uncertainty about the future.

Where long-term stability and security are under threat, such as through for example temporary accommodation, assured shorthold tenancy, risk of eviction and fear of being made homeless, the 'home' becomes something different. If affects the households' ability to participate in day-to-day things that make up so much of our identity. In a study by Crisis and Shelter (2014) participants living in the PRS reported their desire to have a sense of home with security and decent conditions as well as control over their environment to be able to plan for the future. In the majority of cases, it was found that these needs were not being met in PRS.

HMO tenants may have to share space with strangers who may regularly move in and out, representing loss of control over their living environment. Feelings of non–home-like environment may be increased by the provision of some necessary fire safety measures, including fire doors, signage, alarms and call points. There is very little research in this area, although Barratt and Green (2017) explore how HMO tenants create their sense of home, what constrains this and how HMO living might affect identity and well-being. They asked HMO tenants to take photographs of their accommodation and invited comment, finding a range of responses from some tenants with a sense that HMOs were not considered 'proper' places to live due to lack of choice, and there was a feeling of lack of control over shared spaces, and this was aggravated by the temporary nature of the accommodation and marginalisation.

The major cause of homelessness in England is now eviction from the PRS, either retaliatory or the ending of AST. In 2016/17, 18,270 (31 per cent) of housing applications from homeless households cited the ending of AST as their reason for homelessness, and service of notice should trigger the LA's prevention duty under the Homelessness Prevention Act 2017 (Wilson, 2017, 2018). Barriers to homeless and vulnerable people in the PRS include affordability, landlords' reluctance to accept housing benefit recipients, lack of secure tenure, poor housing quality and lack of support for vulnerable people (Cromarty et al., 2017).

The increasing number of households placed in temporary accommodation continues. These 78,930 households include 120,510 children, a 75 per cent increase since 2010. There were 2,030 families with dependent children in B&B–type accommodation at the end of December 2017. LAs have tried to source better accommodation, including outside of their areas, and a 201 percent increase in such households from December 2011 to 2017, with substantial implications for feelings of 'home', the situation being particularly acute in London (Wilson and Barton, 2018). Many are now relocated and rehoused in unfamiliar places, with implications for meaning of home, and there is little research on the effects of this to tenants.

Despite the clear meaning and importance of home, there is currently no requirement to think of the house as 'home' in policy or practice, an area which has received very little attention. Some advocate increased integration of the term 'home', as it would help is think about problems differently, most particularly the human relationship with home and in legal responses to issues identified: there are human, social and personal costs of displacement and dispossession, and a better understanding of home could help enhance interventions and solutions (Fox, 2003; Fox O'Mahoney, 2013). Fox (2003) also raised the fundamental issue that the property is 'home' to the occupiers but 'non-home' to those with commercial interests only, resulting in conflict in decision-making between these interests, arguing that values around the home should have greater status in making legal decisions.

Home: geographical and social life course contexts

Home also comes under close scrutiny when we consider the place setting, particularly when traditional industries and sources of employment are lost and what this means for the local economy and community. Some once-thriving seaside towns such as Margate, Jaywick Sands and Blackpool, which have lost much of their tourist industry, have seen hotels and holiday accommodation converted into permanent rented living accommodation (see, for example, Stewart and Lynch, 2018a). The socio-economic decline and combination of a transient population in more deprived seaside towns has substantial implications around the meaning of home and how local authorities are able to understand and respond to these challenging issues (Barratt et al., 2012; Stewart et al., 2013). See Figure 2.1.

Margate and Jaywick Sands are both cited as examples of housing-led deprivation, which occurs when there is a damaged and dysfunctional housing market; complex, unconnected legislation; an unattractive market for new housing; and lack of ordinary market housing building and housing investment (Elphicke, 2017). Some seaside towns have retained more of a tourist industry, such as Brighton, and have fared differently, although many still suffer pockets of poor housing, homelessness and sense of home. One organisation in Brighton has worked to helped tenants through their Room for Home initiative as part of a process of working with vulnerable tenants and helping create a sense of home. However valuable these projects are, they are volunteer led and subject to resources.

Figure 2.1 Boarded-up housing in a seaside town

Housing needs of PRS tenants change across the life course and require tailored interventions. Rhodes and Rugg's (2018) vulnerability report provides analysis of low-income households in the PRS across the life course. Using evidence they demonstrate that large proportions of tenants are experiencing harm from overcrowding, poverty and poor conditions across the life course, including households with dependent children and households with a person aged 65 or over.

With families and children at risk from poor housing conditions and the Marmot Review arguing that children should be given the best possible start in life, it is important housing is prioritised in strategies for preventing health inequalities in England. There is clear evidence to demonstrate the negative effects of overcrowding on a child's social and educational development. Harker (2006) demonstrates that insecurity has impacts on mental health and well-being including disturbed sleep and stigmatisation leading to bullying. There are also physical effects with higher number of accidents reported in this group and increased exposure to infectious disease. One of the biggest cause of young people's homelessness is family conflict; arguably those affected need greater security of tenure.

'Generation Rent' categorises young people living in the PRS unable to access homeownership or social housing, living in PRS temporarily until they 'settle down' into another tenure and can then benefit from positive qualities of having a permanent home (Hoolachan et al., 2016). As McKee et al. (2017) report, this gives Generation Rent more limited rights than they would otherwise have; it affects their lives and social-spatial inequality. They describe this 'fallacy of choice' as unachievable due to lack of material resources and local housing opportunities.

Saunders (2007) demonstrates that home becomes more important to people as they grow older. There is a growing recognition of the needs of older PRS tenants around lack of control and agency as well as affordability with a fixed or declining income. As a person ages housing conditions may decline alongside their health, requiring home adaptations, and this may require permission from the landlord (Age UK London; Rhodes and Rugg, 2018; Stewart, 2021).

Stewart and Lynch's (2018b) presentation at the Housing Studies Association on PRS and the meaning of home was well received, as was their book of the same year and they have continued to develop their work in this area. Others have also more recently developed this issue and explored the role of PRS landlords and letting agents in helping create a home for tenants (McKee et al., 2021), although based on the Scottish and not the English system.

Conclusions

The PRS can make a positive contribution to housing stock and sense of home when it is affordable, secure and with appropriate conditions and management. However, the PRS is highly differentiated and occupied by more families and older people than in the past. A key policy priority has to be increasing security of tenure so that tenants have a greater sense of control over their homes. The very real power imbalance means leaving tenants in a precarious situation and facing stark choices about whether to try to secure repairs or potentially risk losing their home and having a rental increase. Affordability, conditions and management are also priorities, and policy needs to recognise these as interrelated issues that make up the PRS market.

References

Age UK London. (2018). *Living in fear: Experiences of older private renters living in London.* London: Age UK and Nationwide Foundation.

Barratt, C. and Green, G. (2017). Making a house in multiple occupation a home: Using visual ethnography to explore issues of identity and well-being in the experience of creating a home amongst HMO tenants. *Sociological Research Online*, 22(1), p. 9.

Barratt, C., Green, G., Speed, E., and Price, P. (2012). *Understanding the relationship between mental health and housing in a seaside town*. Colchester: University of Essex.

Barton, C. and Kenny, C. (2018). *Health in private-rented housing*. Number 573 April 2018, Houses of Parliament, Parliamentary Office of Science and Technology (POST) Research Briefing.

Clapham, D. (2010). Happiness, well-being and housing policy. *Policy and Politics*, 38(2), pp. 253–267.

Crisis and Shelter. (2014). *A roof over my head sustain: A longitudinal study of housing outcomes and wellbeing in private rented accommodation (final report)*. London: Crisis and Shelter.

Cromarty, H., Wilson, W., and Bellis, A. (2017). *Private renting solutions for homeless and vulnerable people*. House of Commons Library Research Briefing Number CDP-2017–0043, 6 February 2017.

Easthope, H. (2004). A place called home. *Housing, Theory and Society*, 21(3), pp. 128–138.

Easthope, H. (2014). Making a rental property home. *Housing Studies*, 29(5), pp. 579–596.

Elphicke, N. (2017). *Turning the Tide: Building a faster, stronger, new coastal renaissance*. Housing and Finance Institute.

Fox O'Mahoney, L. (2013). The meaning of home: From theory to practice. *International Journal of Law in the Built Environment*, 5(2), pp. 156–171.

Fox, L. (2003). The meaning of home: A chimerical concept or a legal challenge? *Journal of Law and Society*, 29(4), pp. 580–610.

Harker, L. (2006). *Chance of a lifetime: The impact of bad housing on children's lives*. London: Shelter.

Hiscock, R., Kearns, A., MacIntyre, S., and Ellaway, A. (2010). Ontological security and psycho-social benefits from the home: Qualitative evidence on issues of tenure. *Housing, Theory and Society*, 18(1–2), pp. 50–66.

Hoolachan, J., McKee, K., Moore, T., and Soaita, A.M. (2016). 'Generation rent' and the ability to 'settle down': Economic and geographical variation in young people's housing transitions. *Journal of Youth Studies*, 20(1), pp. 63–78.

Hulse, K. and Haffner, M. (2014). Security and rental housing: New perspectives. *Housing Studies*, 29(5), pp. 573–578.

McKee, K., Moore, T., Soiata, A., and Crawford, J. (2017). 'Generation rent' and the fallacy of choice. *International Journal of Urban and Regional Research*, 42(2), pp. 318–333.

McKee, K., Rolfe, S., Feather, J., Simcock, T., and Hoolachan, J. (2021). *Making a home in the private rented sector: An evidence review*. University of Stirling, Edge Hill, Cardiff University and Safe Deposits Scotland.

Rhodes, D. and Rugg, J. (2018). *Vulnerability amongst low-income, households in the private rented sector in England*. York: University of York, Centre for Housing Policy and Nationwide Foundation.

Saunders, P. (2007). The meaning of 'home' in contemporary English culture. *Housing Studies*, 4(3), pp. 177–192.

Sixsmith, J. (1986). The meaning of home: An exploratory study of environmental experience. *Journal of Environmental Psychology*, 6(4), pp. 2810298.

Smith, M., Albanese, F., and Truder, J. (2014). *A roof over my head: The final report of the Sustain project (Sustain: A longitudinal study of housing outcomes and wellbeing in private rented accommodation)*. London: Crisis and Shelter.

Somerville, P. (1997). The social construction of home. Journal of Architectural and Planning Research, 14(3), pp. 226–245.

Stewart, J. (2021). Meeting the private sector housing condition and adaptation needs of older people: responses from London's environmental health and allied services. *Housing, Care and Support*, ahead of print https://doi.org/10.1108/HCS-03-2021-0009

Stewart, J. and Lynch, Z. (2018a). *Environmental health and housing: Issues in public health*. Oxon: Routledge.

Stewart, J. and Lynch, Z. (2018b). *Environmental health practitioners and the private rented sector: 'House' or 'home' in regulation and public health?* Housing Studies Association Conference: Professionalism, Policy and Practice: Exploring the Relationship between Theory and Practice in Housing Studies, 11–13 April 2018, Sheffield University.

Stewart, J., Rhoden, M., Knight, A., Mehmet, N., and Baxter, L. (2013). Beside the seaside: Perceptions from the 'front line' on the support needs of families living in the private rented sector in Margate. *Journal of Environmental Health Research*, 13(1), pp. 22–23.

Wilson, W. (2017). *Retaliatory eviction in the private rented sector*. House of Commons Briefing Paper, Number 7015, 13 June 2017.

Wilson, W. (2018). *Applying as homeless from an assured shorthold tenancy (England)*. Commons Briefing papers SN06856 Published Monday, June 4, 2018.

Wilson, W. and Barton, C. (2018). *Households in temporary accommodation (England)*. Commons Briefing papers SN02110.

3 Perspectives on the regulatory framework and intervention

Jill Stewart and Zena Lynch

Introduction

In this chapter we seek to unravel the complex regulatory roles in the PRS. We consider the socio-legal framework and some theoretical perspectives on how EHPs operate at the front line of service delivery from persuasion to more punitive approaches. We overview the suite of laws and regulations and balancing tensions involved in delivering better conditions in the PRS and – more individually – the health, safety and security of a tenant's home.

EHPs at the front line of service delivery

The EHP governance role in regulating the PRS is wide and varied, utilising a socio-legal framework, and can be considered through a range of theoretical perspectives as indicated in Figure 3.1. It encompasses assessment, correction, control and prevention of environmental stressors. Despite a long history of intervention, little is written about how EHPs regulate the PRS, and most critiques are around the food and health and safety aspects of their work. There are, however, some parallel studies that relate to front-line practitioners with similar public service roles and how they use their discretion in the complex, sometimes confounding day-to-day aspects of applying regulations. Different LAs and practitioners may have different values, even moralities, in delivering and administering their statutory and discretionary functions, and there may be variation at the front line of delivery. The spectrum of intervention may range from strict enforcement and application of law on the one hand to a more client- or person-centred approach by another practitioner, and EHPs have immense discretion in delivering their roles.

Lipsky's (1980) seminal theory on 'street-level bureaucrats' concerns the way in which front-line practitioners deliver public policy and practice, particularly at times of austerity. Lipsky suggests that conflicted street-level

DOI: 10.1201/9781003246534-4

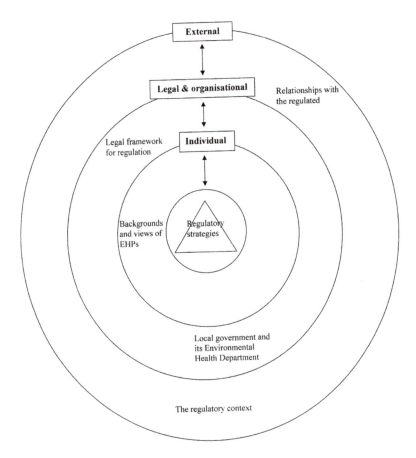

Figure 3.1 The governance model of regulation by local government EHPs

Source: Reproduced with kind permission from Couch (2016)

bureaucrats essentially become policy makers as they interact with citizens and 'process' people through the system. Lipsky considers how people experience public policies, arguing that it is not entirely top down but created within the crowded offices and daily encounters of the street-level bureaucrats. They essentially become the policies they are carrying out, and the work becomes a cycle of policy conflict and struggles between individual workers and citizens, who both challenge and submit to being processed as clients.

Many EHPs joined the profession and prioritised the PRS aspect of their work with a sense of social purpose and justice but find that they are unable to do what they set out to. What they found – Lipsky supposes – is inadequate resources, unpredictable clients, a lack of training in meeting new and workloads prohibiting individualised practice, leading to 'mass processing' of clients. Street-level bureaucrats learn to cope by compromise within the limits imposed. More recent studies suggest the attitude of EHPs themselves can contribute an inflexible approach, with a reluctance to move away from a 'health inspector' model in favour of community development approaches (Dhesi and Lynch, 2016). The clients' experience of public policy may not be what they were seeking.

The demand for EHPs' work exceeds the resources they have available to deliver basic statutory work in inspection and enforcement before even factoring in a more person-centred approach. Lipsky calls this a form of surrogate in effort to appear accountable through performance targets, yet there are no indicators around the quality of inspection or follow-up. We also know that enforcement levels remain low in this sector despite statutory obligations on local authorities to act (Battersby, 2015). This is due to a range of issues, including difficulties in recruiting suitably qualified EHPs as well as lengthy regulatory/enforcement processes, with varying degrees of effectiveness.

Lipsky argues that interventions should be more client centred, with citizens inputting into decision-making around things that affect them. Housing cases can be fraught with complexity and uncertainty for a range of reasons. Developing this understanding of public services further, Maynard-Moody and Musheno (2003) analyse stories of front-line practitioners which can help us to understand how EHPs give identities to their clients in order to account for their own behaviours, decisions and regulatory interventions. Again, this can help us make sense of EHP decisions and judgements at the front line of practice and the manner in which they intervene to support or to sanction clients.

EHPs are seen to act 'reasonably' and apply discretion in their approach to law enforcement, taking a legal and moral stance within a social environment in their work and using a range of approaches to deliver regulation (Hutter, 1988). Hutter also suggests that EHPs use both persuasive and insistence strategies as part of an accommodative or compliance model. Persuasion remains dominant, and the development of social relationships between the regulated and those they control is primary in determining the enforcement approach adopted. Since publication of this seminal work, practically all elements of law and practice around PRS housing conditions and management have changed, and there has been more of an emphasis on enforcement using a range of civil and criminal remedies. We are therefore seeing a situation where EHPs operate within law which must be interpretive and understanding of the social environment and structural process within which it occurs, and ultimately there are criminal sanctions.

Developing this work, Hutter (1997) purports that theories of regulation divide around whom the government intervene on behalf of, and of course housing law and regulations should be about *protecting* the tenant, i.e. the less powerful party. Conflict theorists would see economic interest as paramount and recognise the tensions between conflicting groups (landlord/tenant), and the resolution of these tensions is viewed in different ways. They see that economic interests are generally protected and that interests are not seriously affected by legislation. In contrast accommodative theorists argue this is due to a general low level of enforcement due to limited resources, ambiguous legislation and weak sanctions enforcement. Consensual theorists, in contrast, suggest that laws have improved but not on the scale needed (as cited in Hutter, 1997). Following the standard pyramid model of responsive regulation (Hutter, 1997) also assumes that those regulated will behave rationally and respond to increasingly punitive approaches, but this remains a little-researched area.

More recently arguments around regulation in environmental health have been critiqued in relation to effectiveness of those charged with regulation and enforcement and the burden on business and 'red tape' arguments. With an overarching policy drive toward 'better regulation' targeted on higher-risk premises and those displaying less compliance, Tombs (2016) had researched environmental health regulation around food and health and safety delivered at local authority level. The principles are also relevant to PRS, since many of the same pressures of a trend toward cutting red tape are felt by private sector housing enforcement teams.

The general reduction in PRS regulation in itself is complex and due to austerity cuts, outsourcing, forms of self- (or no) regulation, and Tombs (2016) sees this a decline in the Las' ability to intervene in social protection. These pressures are felt in PRS enforcement as staff recruitment and retention remain problematic and due to the exponential increase and complexity of the PRS itself (see also Battersby, 2015). There are challenges to address problems in the PRS particularly, requiring a strategic, competent and skilled multi-agency approach in enforcement of conditions and management as well as tenancy relations and security. Tombs argues that we notice the lack of enforcement only when things go wrong, although this is of course only part of the picture, since there is not always a linear relationship between cause and effect; it is usually far more complex.

The need for robust data is clear in the ability to deliver more proactive and area-based approaches to the PRS, which arguably could drive out the worst landlords for the market, bringing about more stability in the local environment and community and sense of home for tenants. A more proactive and nuanced approach could contribute to a trend over time of a better regulated PRS. The problem is that much of this evidence is unpublished, and there needs to be more rigorous analysis of trends in data over time to assess effectiveness.

Laws and regulations governing the PRS in England

The main statutes governing housing conditions in England include the Housing Act 2004 and the Housing and Planning Act 2016, supported by a range of regulations, as illustrated in Table 3.1. However, it is more complex than just the law; it is also a question of the rapid growth and differentiation of the sector, austerity and loss of experienced EHP workforce (Battersby, 2018) and challenges of strategic approaches in LA work in enforcement, planning and homelessness (Rugg, 2020). Addressing housing, health and home in the PRS remains challenging and has become the subject of increasing interest at parliamentary level (see, for example, Barton and Kenny, 2018). Despite all this dynamic legislation, many 'homes' remain in poor condition, and there is complexity and some divided opinion and legislation on how we should be moving forward with different perspectives around regulation versus too much bureaucracy.

Protection from Eviction Act 1977	Protection from eviction without a court order, with exceptions
Housing Act 1985 (as amended)	Various provisions still apply (e.g. overcrowding), many now superseded elsewhere
Housing Act 2004	Ongoing requirement to keep housing conditions in area under review and identify action to take
	Introduces HHSRS and enforcement procedures, mandatory and discretionary licencing for parts of PRS, mainly HMOs (Additional Licencing) but also the wider PRS (Selective Licencing) and management orders, new Residential Property Tribunals and enforcement protocol
	Supported by ODPM (2006) Operating and Enforcement Guidance (statutory) – currently under review (see Chapter 8)
The Management of Houses in Multiple Occupation Regulations 2006	Covers landlord and tenant requirements for standards of management in HMOs
Local Government and Public Involvement Act 2007; Health and Social Care Act 2012 and Care Act 2014	Duty for health and LA to work together to produce Joint Strategic Needs Assessment (JSNA) (see Chapter 4) of health and well-being of local community, with an emphasis on prevention (includes PRS housing)

Housing and Planning Act 2016	New provisions for Banning Orders, Rogue Landlord database, fixed penalties and extended provisions for Rent Repayment Orders
Deregulation Act 2015	Remedy for retaliatory eviction
Energy Efficiency (Private Rented Property) England and Wales Regulations 2015	Require defined poor-energy-efficient PRS properties to be improved
The Licensing of Houses in Multiple Occupation (Prescribed Description) (England) Order 2018	Covers premises defined as HMOs and criteria for this
The Licensing of Houses in Multiple Occupation (Mandatory Conditions of Licences) (England) Regulations 2018	Conditions relating to licencing and room sizes used for sleeping accommodation
Homelessness Reduction Act 2017	Duties to those homeless or threatened with homelessness
Homes (Fitness for Human Habitation) Act 2018	Civil remedy consolidating and updating earlier law; implies fitness for human habitation during tenancy
Electrical Safety Standards in the Private Rented Sector (England) Regulations 2020	Landlords must ensure fixed electrical installations are inspected and tested at least every five years by a qualified person

The PRS rented sector is covered by a huge raft of laws and regulations about property conditions and management but also around the tenancies of those who live there (Rugg and Rhodes, 2010). Despite this, there are still some gaps in coverage, also aggravated by unbalanced power relationships between landlord and tenant (Stewart, 2007). These laws are complex to apply reactively to housing stock that can be of very poor or sometimes inappropriate design or conditions not really suited to purpose and poorly managed, and there are differences in adoption of discretionary powers nationally. As new powers such as for Rent Repayment Orders and Banning Orders have been rolled out, local authorities are adopting their own local policy and practice interventions, which are variable (see Chapter 7).

Another layer in the complexity is the difference between mandatory and discretionary powers, particularly with property licencing. Not all properties require a licence; only houses in multiple occupation (HMO – five or more residents forming at least two households) require a mandatory licence. Every LA has a different approach to discretionary licencing, which may be for Additional HMOs or Selective Licensing across the general PRS for all or part of their local authority area. Licencing can help to identify the worst landlords and seek to drive them out of the market, since only a 'fit and proper person' can hold a licence. Overall though the situation is erratic and mixed nationally. See Chapter 10.

Chartered Institute of Environmental Health (CIEH) is now calling for a national landlord registration scheme in England (Matthewman, 2019). Other UK nations already have registration schemes embedded as policy, helping to identify and collate information on landlords and agents. Ireland's mandatory landlord registration scheme collates information about the PRS and they have longer-term tenancies than is generally the case in England; the combination is found to particularly support low-income tenants (Moore and Dunning, 2017). Registration is different than licencing, which is more about compliance of property management and the wider geographical area, such as low demand and anti-social behaviour. Despite having mandatory and discretionary licencing in England there are major gaps in coverage.

Conclusion

EHPs operate within a socio-legal framework and require multiple skills and competencies in delivering better conditions and management in the PRS and to help ensure security for tenants where they can. Laws and regulations are dynamic and require continued investment in staff training from Las. At a time of such growth, complexity and challenges to regulation of the PRS Las would benefit from developing procedures and activities to better manage the PRS and coordinating the wider range of enforcement powers. Surely alongside this, further consideration could be given to thinking about home if the PRS is to provide a better 'product' in the marketplace. There needs to be more flexibility and reflexivity around regulatory goals, not just conformity or not, which fails to address a wider sense of home.

References

Barton, C. and Kenny, C. (2018). *Health in private-rented housing, number 573 April 2018, Houses of Parliament, Parliamentary Office of Science and Technology (POST) research briefing*. London: HoC.

Battersby, S. (2018). *Private rented sector inspections and local housing authority staffing supplementary report for Karen Buck MP*. London.

Battersby, S.A. (2015). *The challenge of tackling unsafe and unhealthy housing: Report of a survey of local authorities for Karen Buck MP*. London.

Couch, R. (2016). *Environmental health regulation in urban South Africa: A case study of the Environmental Health Practitioners of the City of Johannesburg Metropolitan Municipality*. Unpublished PhD Thesis, London South Bank University.

Dhesi, S. and Lynch, Z. (2016). What next for environmental health? *Perspectives in Public Health*, 136(4), pp. 225–230.

Hutter, B.M. (1988). *The reasonable arm of the law: The law enforcement practices of environmental health officers*. Oxford: Clarendon Press.

Hutter, B.M. (1997). *Compliance: Regulation and environment*. Oxford: Clarendon Press.

Lipsky, M. (1980). *Street-level bureaucracy: Dilemmas of the individual in the public services*. New York: Russell Sage Foundation.

Matthewman, R. (2019). *Press release: CIEH calls for national landlord registration scheme*. Available: www.cieh.org/news/press-releases/2019/cieh-calls-for-national-landlord-registration-scheme/

Maynard-Moody, S. and Musheno, M. (2003). *Cops, teachers, counselors: Stories from the front lines of public service*. Michigan: University of Michigan Press.

Moore, T. and Dunning, R. (2017). *Regulation of the private rented sector in England using lessons from Ireland*. York: Joseph Rowntree Foundation. Online. Available: www.jrf.org.uk/report/regulation-private-rented-sector-england-using-lessons-ireland (accessed 4 March 2020).

Rugg, J. (2020). *London Boroughs' management of the private rented sector*. London: Trust for London.

Rugg, J. and Rhodes, D. (2010). 'Between a rock and a hard place': The failure to agree on regulation for the private rented sector in England. *Housing Studies*, 18(6), pp. 937–946.

Stewart, A. (2007). Rethinking housing law: A contribution to the debate on tenure. *Housing Studies*, 9(2), pp. 263–270.

Tombs, S. (2016). *Better regulation: Better for whom?* London: Centre for Crime and Justice Studies, Briefing 14 April 2016.

4 Partnerships of prevention

Beyond regulation

Ellis Turner

Introduction

The Health and Social Care Act 2012 amended the Local Government and Public Involvement in Health Act 2007 in respect of duties and powers around Health and Wellbeing Boards (HWB), Joint Strategic Needs Assessments and Joint Health and Wellbeing Strategies, supported by statutory guidance (Department of Health, 2012). Dhesi's research (2019) explored how we can reinvent the environmental health role and more effectively tackle health inequalities in the context of public health, particularly in the LA context. Despite ongoing and largely positive moves, anecdotal evidence suggest that few Joint Strategic Needs Assessments (JSNA) focus greatly on the PRS, where some of the worst housing conditions and health inequalities and inequalities experienced across the life-course.

This chapter therefore asks: What are the opportunities for EHPs to work in partnerships with health, social care and the third sector, and how effective are these strategies? What is considered best practice at reducing harm and risk by changing behaviour at scale? Regulatory or partnership interventions? Maybe elements of both? If so, how should these elements be balanced? If our resources are scarce, then evaluating all our interventions to identify the most effective practice could help deploy our limited resources to where they yield the greatest health impact and benefit.

Such a critically reflective line of enquiry across our professional practice can contribute to a wider body of knowledge and develop further the profession's aspirations. Some of the context of the chapter draws from the author's professional practice and partnerships formed with health and social care teams and the third sector when in an NHS public health–funded EHP post. The objectives of this were to nurture partnerships of prevention, use data to inform interventions and promote the role of the EHP across stakeholders. This chapter explores some of the drivers of the modern

DOI: 10.1201/9781003246534-5

public health agenda and the relevance and opportunities of environmental health practice today.

The modern public health agenda – drivers of change

The last decade as seen some significant changes to the organisation of health and social care with some emerging progressive ideas on the need for prevention. More detail on this is available in Stewart and Lynch (2018) together with information on – for example – the Public Health Outcomes Framework as well as evidence-based NICE Guidance and Pathways, which help guide our thinking in how we approach and tackle housing and health inequalities in the PRS. We need to think beyond our immediate (and busy) roles and refocus on where we can most effectively intervene and with whom we should work to deliver better outcomes.

As we live longer the impacts of austerity are experienced more acutely in the ageing population, particularly in older and less adaptable housing stock. It is argued by some that perhaps our efforts would be best spent invested with vulnerable clients to help mitigate the severity and likelihood of long-term conditions? Initiatives such as the NHS 5 Year Forward attempted to express the need to invest in health prevention services now to avoid health care services becoming financially unsustainable in the future. One intervention opportunity that was presented to EHPs was the 'making every contact count agenda', whereby EHPs and other health practitioners refer vulnerable clients into a single point of access to health assessment hubs such as the Seasonal Health Intervention Network (SHINE), who could then be signposted into a range of prevention services such as clinically commissioned health navigator services or housing adaptions support.

The Care Act 2014 also presented new opportunities for EHPs bringing back public health and social care teams together to work closer and in a more multidisciplinary strategic way, such as developing health mapping to express priority areas of need or client-centred services such as supporting hoarding behaviour. Marmot (2020) suggests that the gap of health inequalities has widened in the last decade, so the strategy of working across all the social determinants to tackle health inequalities (Marmot, 2010) is even more relevant today, with critical literature available to inform opportunities for EHPs to work in effective partnerships of prevention.

Beyond the individual – wider interventions

Some health practitioners argue that left to individual choices, it is a challenge to mitigate health hazards and more challenging the steeper the social gradient. For example, as inequalities from limited choices increasingly prevail, such as poverty, limited access to safe employment, good education, healthy food and homes, so does the steepness of the gradient. The introduction of social and environmental interventions can reduce this gradient by enabling equitable access to high-quality health services and amenities, which in turn can empower communities and individuals. Such interventions mitigate health hazards (see, for example, Figure 4.1), increase healthy

Figure 4.1 Damp and mouldy bathroom

spaces, reduce inequalities and can improve well-being. So where do EHPs start to nurture these partnerships and express their health prevention value?

Health policy priorities and development of local health programmes

One key place to start developing networks and relationships is with key decision makers and teams that inform the development of local health policy to ensure the preventive value of EHPs is fully understood and their potential is realised. Evaluation of interventions such as hazard mitigation can be critical in making the business case e.g. using the Building Research Establishment (BRE) Housing Health cost calculator (see www.housinghealthcosts. org/) to quantify the savings made to NHS, society and number of persons no longer requiring health treatment (see also Chapter 12).

Other strategies include making subtle differences to our language when communicating with health providers and clinical commissioning groups. E.g. EHPs can be described as primary preventers of ill health who work across the social determinants of health to prevent accidents and infections and ensure physiological and psychological requirements are met. They do this by identifying, assessing and managing risk through regulation or partnerships.

Advocacy and representation at local health policy-making groups is critical to ensure housing is recognised as a determinant of health that can inform other determinants of health. EHPs need to ensure that they maximize all opportunities to promote their housing and health work in developing JSNAs to continue gaining traction in their critical work in reduce health inequalities.

Whole-systems approach – integration of risk factors

The relative contributions of a range of health factors to selected health outcomes can be expressed: for example, the quality (or risk of harm) from our diet, employment, occupation, community safety, homes and wider environment can greatly inform our life chances up to 40% (Park et al., 2015). Many of these risk factors are disciplines of environmental health practice. A concise interpretation of Dahlgren and Whitehead's determinants of health and well-being in our neighbourhood's conceptual model is who you are, what you do and where you live.

Accurate public health data – health mapping by EHPs to express needs

Using data can help inform and develop wider health interventions that seed to improve the quality and quantity of affordable housing provision to meet expressed housing needs and health priorities.

Local Public Health Outcomes Framework data sets can identify a range of health indicators that has a relationship with housing and neighbourhoods. More focused searches can identify health indicators with a more direct relationship with housing. Analysis and review of this data and local-authority health profiles can inform a view to what are the local health priorities. This can be the start of conversation with commissioners and health development teams about what wider housing interventions could mitigate local health inequalities.

The CIEH health mapping tool kit (Hunter and Buckley, 2013) sets out a comprehensive guide as to how EHPs can develop these tools further to express the health needs of an area and respond to the public health agenda. Since 2013 various collaborations by EHPs have been pioneered to map out the local needs of an area to inform a wider housing intervention either in partnership with health, social care and public health teams or regulatory interventions such as property licensing schemes. Early public health pioneers such as Booth and Snow started this hundreds of years ago (Stewart, 2017), and the art and science of environmental health mapping continues today.

Whilst a practitioner, this author collaborated with public health team colleagues to create a health map that could express a health need and potentially a housing intervention. The local health profile needs of the London Borough of Islington identified that hip fractures in the over-65 population was above average. We knew that Islington had one of the highest rates of injuries from falls in England. In 2011 Alex Singleton, a lecturer in geographic information science (GIS) at Liverpool University, downloaded every census dataset for every local authority in England and produced a free library of downloadable PDFs. Using Alex Singleton's analysis of the 2011 census data we simply overlaid the data of where the PRS was concentrated in the borough and GPs surgeries where over-65s with fractured hips had been registered.

Was there a relationship between housing stock and falls? A map was created indicating where proactive street surveys could be focussed to help mitigate where falls hazards were located. A health prevention case could be made to fund more EHPs to, for example, to target cold and falls hazards and be able to demonstrate the cost benefits. The PRS census data and fall health data are just two data sets from hundreds that could have been mapped. What data sets could you match to express where the needs or health inequalities in your area are, such as heating provision and mental health?

The evolution of predictive analytics to identify elements of the PRS and its needs has evolved considerably in the last decade and is now quite sophisticated. Early partnerships by London office data analytics to support local-authority environmental health teams to identify their houses in multiple occupation (HMOs) using their internal data helped develop this practice and progress big data conversations amongst EHPs. LAs can now share best practice as to how to interrogate council tax or tenancy deposit data to identify their PRS in the absence of borough-wide licencing schemes or a national landlord register. These innovations have been pioneered by EHP-led consultancies that can provide local authority with services to help analyse their data to identify the PRS and housing need such as location of potentially high-risk unknown HMOs and areas potentially most in need of risk assessment and investigation. Once the local PRS has been identified, other data such as anti-social behaviour (ASB) and other housing complaints can be overlaid onto maps to identify expressed needs and health inequalities for proactive interventions.

Partnerships of prevention with EHPs – social prescribing projects and interventions

EHPs have led and developed many wider housing interventions. Following are some examples of the types of partnerships to which EHPs continue to contribute.

Hospital discharge planning services

The National Institute for Health Care Excellence (NICE) excess winter deaths and cold home guidance NG6 (NICE, 2015) could be considered a paradigm shift in collaborative preventive health care. Twelve recommendations were made as to the role other health practitioners could make in preventing recurring readmissions and excess winter deaths caused by cold homes. Historically a health care practitioner may have checked whether there is food available at home for discharged patients, but how about assessing, heating, insulation, hot water and falls? If cross-tenure pathways exist for rapid repair and improvement, then winter deaths and readmissions for falls and so on can be prevented. Only one element of the recommendations made reference to EHPs. A progressive interpretation could argue that EHPs could and perhaps should be involved in many more of these.

There are some models of good practice around hospital discharge; see, for example, Care and Repair's Manchester scheme (Lineacre, 2015). At the London Borough of Islington 2014–2015 an EHP led a collaboration with the Clinical Commissioning Group (CCG) who requested Age UK, floating support and funded posts to assess clients housing needs upon admission and not discharge. Services and pathways to support were enabled to carry out repairs such as handrails to stairs, heating and hot water, and so on. A cross-tenure partnership was set up with housing services and EHPs to ensure that regardless of tenure, a rapid repair could be actioned upon being identified. This included urgent accident prevention and repair grants for vulnerable owner occupiers. Perhaps the most significant evaluation and feedback from clinicians is that these schemes are inclusive and tenure blind and are best served by having a single point of access or referral/assessment point similar to the SHINE single-point-of-access health-assessment hub.

Multidisciplinary clinical teams

One aspiration of EHPs is for all health assessments to include enquiries as to clients' housing conditions. Attempts to support GPs to do this may have limited success, but there are some commissioned services using locality navigators that help support clients with health needs being at the centre of their health services provision. They can signpost clients to a range of services, including those of EHPs, to help enable a healthy home. At Islington in 2014–15 EHPs successfully negotiated articipation in the clinical multidisciplinary teams (MDTs) during GP case conferences. These MDT case conferences consist of a range of health practitioners who try to unpick the causes of ill health from clients with the most complex health and long-term conditions. Other health practitioners including those from the SHINE team and a specialist community social worker with oversight of the relationship with housing and health were also able to promote work around addressing housing conditions and assessments.

Contributions to boiler replacement programme and thermal insulation project evaluation

In 2014 Islington Council secured capital funding to deliver a borough-wide non–means-tested boiler replacement programme. Housing EHPs were well placed to contribute to this programme by identifying throughout their work inefficient F- and G-rated boilers for replacement. Such interventions can mitigate cold hazards, reduce fuel poverty and lower CO_2 emissions.

A similar collaboration involved Islington's EHPs, public health team and housing services team. When external wall insulation had been installed to a block of flats, an evaluation was carried out of the work on the impact to residents' health. EHPs were able to contribute primary data to the evaluation by sharing the information from the housing health cost calculator that they had from assessing conditions before and after the works using the Housing Health and Safety Rating System (HHSRS) (see also Chapter 8). This data was then used to make indicative projections to extrapolate and quantify the health benefits in terms of prevention of illness and treatment costs if improvements had not been carried out.

Supportive hoarding pathways for improved outcomes

There are many examples of good practice of EHPs contributing and developing multidisciplinary team hoarding protocols through partnership with health and social care teams and the third sector (Ousta Doerfel and Jonas, 2015). These partnerships can ensure clients' needs are best met, especially now that hoarding behaviour has a clinical diagnosis, with some treatments available and pioneering research as to the complex cause.

One example of good practice involves hoarding panels similar to domestic violence and anti-social behaviour panels. These cross-service panels with client-led advocacy can enable clients' needs to be expressed and then supported and can also offer support for practitioners to inform meaningful and decision-specific capacity assessments and other supportive pathways. Other EHP-led innovations include hoarding grants for a range of therapies and services to support addressing both the symptoms and causes (Turner and Karnes, 2016). EHPs in partnership with Hoarding UK and the CIEH currently offer support for EHPs managing complex hoarding cases in the form of bespoke training and quarterly drop-in support sessions for EHPs to share and learn best practices to support clients with hoarding behaviour.

Conclusion

In guiding EHP thinking around partnerships of prevention, the following questions are helpful pointers:

- What wider housing interventions could EHPs lead on? What are the health priority inequalities from the local health profile? What are the social determinants of health within the LA that are still not being sufficiently tackled and why? How do we nurture relationships to increase the collaborative appetite of stakeholders? How do we overcome the challenges of political buy-in? Do we need to make the economic business case, the health prevention case or both cases?
- What partnerships can be developed and with whom? Who are the key decision makers in your health, social care and third-sector teams? How can EHPs inform their decision-making for local health programme development and commissioned services?
- Are housing needs addressed or conditions recognised as contributing to ill health in your joint strategic needs assessment? If not, why not,

and what role could EHPs contribute? How can we specifically focus on the PRS?

- Is the relationship with housing and health expressed as an area of priority need with the health and well-being board? Is housing represented in the makeup of the board? What do EHPs do to influence priorities or policy thinking?
- What could be health mapped? Are EHPs in conversation with the local public health team? And what is already mapped that can help develop a better understanding of the PRS?
- Consider reviewing the local health profile. What local Public Health Outcome Framework indicators could be mapped against one of the census categories? What could this demonstrate? And what potential partnership intervention could this help inform?

Both the PRS regulatory framework and the public health agenda offer a range of opportunities to help focus on delivering more effective interventions, and ongoing partnership working can help ensure that services are targeted towards reducing housing and health inequalities and inequities at their most acute.

References

Department of Health (DoH). (2012). *Statutory guidance on joint strategic needs assessments and joint health and wellbeing strategies*. London: DOH.

Dhesi, S. (2019). *Tackling health inequalities: Reinventing the role of environmental health*. London: Routledge Focus.

Hunter, J. and Buckley, S. (2013). *Mapping health toolkit*. London: CIEH and Nottingham Trent University. Online. Available: www.cieh.org/media/1253/mapping-health-toolkit.pdf (accessed 5 October 2021).

Lineacre, J. (2015). *Manchester care and repair home from hospital service, from Kings Fund: Bringing together housing and public health: Enabling better health and wellbeing conference*. Available: www.kingsfund.org.uk/events/housing-and-public-health and www.kingsfund.org.uk/sites/default/files/media/Janette_Lineacre_Manchester_Care_and_Repair_Home_from_Hospital.pdf (accessed 5 October 2021).

Marmot, M. (2010). *Fair society healthy lives (the marmot review)*. Online. Available: www.instituteofhealthequity.org/resources-reports/fair-society-healthy-lives-the-marmot-review (accessed 5 October 2021).

Marmot, M. (2020). *Health equity in England: The marmot review 10 Years On (2020)*. Online. Available: www.instituteofhealthequity.org/resources-reports/marmot-review-10-years-on (accessed 5 October 2021).

The National Institute for Health and Care Excellence. (2015). *Excess winter deaths and illness and the health risks associated with cold homes NICE guideline [NG6] March 2015*. Online. Available: www.nice.org.uk/guidance/NG6 (accessed 5 October 2021).

Ousta Doerfel, S. and Jonas, A. (2015). *Islington hoarding protocol*. London Borough of Islington. Online. Available: https://uwe-repository.worktribe.com/output/5632370

Park, H., Roubal, A.M., Jovaag, A., Gennuso, K.P., and Catlin, B.B. (2015). Relative contributions of a set of health factors to selected health outcomes. *American Journal of Preventative Medicine, Elsevier*, 49(6), pp. 961–9. Available: www.ajpmonline.org/article/S0749-3797(15)00405-5/abstract

Stewart, J. (ed.). (2017). *Pioneers in public health: Lessons from history*. London: Routledge Focus.

Stewart, J. and Lynch, Z. (2018). *Environmental health and housing* (2nd ed.). Oxon: Routledge.

Turner, E. and Karnes, M. (2016). *Hoarding behaviour fact sheet*. London Borough of Islington. Online. Available: https://uwe-repository.worktribe.com/output/912966

Useful websites

Building Research Establishment (BRE) – Housing Health Cost Calculator available as subscription service: www.housinghealthcosts.org/

Islington Seasonal Health Interventions Network (SHINE) – www.energyadvice.islington.gov.uk/energy-advice-team/shine/

Local authority health profiles – https://fingertips.phe.org.uk/profile/health-profiles

The National Institute for Health and Care Excellence (NICE) – www.evidence.nhs.uk/ – for Evidence Pathways and Guidance, as well as open access resources

Public Health Outcomes Framework – https://fingertips.phe.org.uk/profile/public-health-outcomes-framework and https://fingertips.phe.org.uk/search/housing

5 Practical problems before the courts and tribunals

David Smith

Introduction

This chapter focuses on a range of more strategic issues associated with enforcement activity. It discusses the different courts and tribunals that are involved in housing work and some of the issues that need to be considered when creating an integrated housing enforcement policy.

Understanding the courts and tribunals

England and Wales have a fairly complex court structure. This has been made more complex in the housing sector by the addition of a range of tribunals which deal with key housing matters.

The courts

The courts are split into two streams, the civil and the criminal courts. Even within these streams there are overlaps between them, and so the magistrates deal with matters that might appear to be civil in nature, while the civil courts often deal with matters that have the appearance of being criminal. So, for example, the magistrates deal with complaints by tenants or LA officers about landlords' properties under the Environmental Protection Act 1990. Housing repairs are usually thought of as more of a civil matter, and the magistrates can award tenants damages for reaches by landlords. At the same time the County Court, a civil court, deals with penalties for failure to protect a tenant's deposit, even though those penalties are more than tenants' actual losses and so have a penal aspect to them.

The Civil Courts are made up of: .

The County Court – The County Court was originally added to take pressure off the High Court by dealing with smaller civil matters but the

DOI: 10.1201/9781003246534-6

size and range of matters it deals with has grown over time. In the housing world the main cases found in the County Court are landlord possession claims and claims by tenants for disrepair under the Landlord and Tenant Act 1985.

The High Court – This is the original court of England and Wales. It is particularly important, as it is invested with a level of authority that stretches beyond specific laws made by Parliament and has a general role as a justice-giving body which has a constitutional significance. High Court judges are appointed by the Crown and serve at the Crown's pleasure, and this is supposed to signify their substantial independence from the other parts of the state. Appeals from the County Court usually go to the High Court, and High Court decisions will bind judges in the County Court, who will have to follow those decisions by interpreting the law in the same way.

The Court of Appeal (Civil Division) – This is the main civil appeal court in England and Wales. Some County Court appeals go straight to the Court of Appeal if they have already been appealed in the County Court, and appeals from High Court decisions come to the Court of Appeal. Court of Appeal decisions will in most cases represent the last word on the law in an area.

The Supreme Court – This is actually the newest court in the UK legal landscape. Previously appeals from the Court of Appeal went to the House of Lords and were actually heard by a group of judges sitting in the House. As part of a process of tidying up the UK constitutional position during devolution a new court, the Supreme Court was created to be the final court of appeal in the UK. It hears both civil and criminal matters but hears a relatively small number of cases. Very few housing cases end up here, but those that do are likely to involve very complex issues of law with far-reaching consequences.

The Criminal Courts are made up of:

The Magistrates Courts – All criminal matters start in the Magistrates. A lot of smaller matters are dealt with exclusively by the Magistrates, but bigger matters also pass through their hands on the way to the more senior courts.

The Crown Court – The Crown Court is the court which represents what most people think of as a criminal trial. In the Crown Court cases are tried before a jury in front of a judge. This is reserved for the most serious cases, and few housing matters would be found here. Appeals on the facts from the Magistrates Court also come to the Crown Court to be heard by Crown Court judges without a jury.

The High Court – The High Court also has a role in criminal cases. Sitting as the Divisional Court, it deals with appeals on points of law from the Magistrates courts. Having decided the law, the matter is sent back to the Magistrates to make a decision in the light of the guidance from the High Court.

The Court of Appeal (Criminal Division) – The Court of Appeal is also important in criminal matters, but it is a totally separate division from the civil division, and judges from one division will not normally deal with cases from the other. Only the most serious and important criminal cases will find themselves before the Court of Appeal, and in practice, such appeals rarely succeed.

The Supreme Court – The Supreme Court is the ultimate route of criminal appeal in the UK. Very few cases will come before the Supreme Court from the criminal courts, but those that do will often lead to substantial changes in the course of the criminal law.

The tribunals

The tribunals are very different from the courts, but they have an increasing importance in housing matters.

The tribunals are organised into First-Tier and Upper Tribunals. Almost all matters start in the First-Tier Tribunal, with the Upper Tribunal handling appeals from the First Tier. Beyond the Upper Tribunal appeals go to the Court of Appeal (Civil Division) and then, in very rare cases, the Supreme Court. The First-Tier and Upper Tribunals are further sub-divided into chambers which deal with different areas of law. For housing matters the key chamber is the property chamber, which deals with appeals under the Housing Act 2004 and a range of other connected property issues. The other chamber of note is the general regulatory chamber which has a role in dealing with appeals from letting and estate agents over penalties for breach of their regulatory regimes. Each chamber has its own rules and procedures, which are quite different.

It is important to realise that a Tribunal Is a creature of statute. That means that it can only do those things that it is specifically authorised to do by statute. It cannot give itself additional powers or have those powers given to it by anyone else (*Avon Ground Rents Ltd v Child*). This is of particular importance in relation to the Housing Act 2004. For example, there is no mechanism to resolve disputes as to whether or not a property is an HMO. The only option is for a LA to prosecute or issue a penalty and for it to be contested before a court or tribunal on that basis (*London Borough of Islington v The Unite Group Plc*).

The other key point to consider is the nature of the hearing. Most court hearings against LA decisions are appeals, which means that the court considers the decision coming before it and whether it was right. On appeal, a decision is only likely to be altered if it is unlawful or seriously wrong; a substantial margin of discretion will be allowed to the original decision maker. Most tribunal hearings, by contrast, are re-hearings. This means that the tribunal is substituting its decision for that of the body being appealed, usually the LA. Therefore the level of respect accorded to the LA decision is much lower and the LA will need to argue before the tribunal for the decision all over again.

Investigating the offence

All investigation should move toward a prosecution. What is meant by this is that any serious investigation should assume the possibility of prosecution or a contested civil penalty hearing. This is not to say that LA officers should assume that every landlord they investigate is guilty; quite the opposite. A LA officer should be a dispassionate investigator and should keep in mind the very real possibility that those they are investigating are either innocent or should not be prosecuted, as their offending is of a minor nature and little purpose will be served by pursuing it. This is not always something that officers remember, and they can at times give the appearance that they have decided on guilt at an early stage and will do anything possible to find it. The key point is that while keeping an open mind it is necessary to ensure that evidence is collected which heads towards proving a specific set of offences.

It is also important to avoid irrelevance. In many cases good LA cases are undermined by being lost under a welter of irrelevant information. This tends to happen because witness statements are produced which partly link to offences which are ultimately not charged and not edited down when those offences are not pursued. This means that they contain information that has no relevance to the offences that remain, which tends to obscure the remaining good evidence.

Offences of strict liability

Almost all the housing offences that are dealt with by LAs tend to be offences of strict liability. This means that Parliament and the courts (*Mohamed & Anor, R (on the application of) v London Borough of Waltham Forest*) have decided that the intent of the offender is if no relevance, and the mere facts of the case are enough. So, for example, a failure to have an HMO licence where one should exist is an offence under s72, Housing Act 2004. The reasons this has occurred are not important, and the fact that the landlord did not intend to make this happen does not matter.

There are some limited defences. So, under s72(4), Housing Act 2004, there is a defence to any form of failure to have an HMO licence if an application has been made for a licence or Temporary Exemption Notice and has yet to be determined.

There is also a defence to many strict liability offences that the offender has a "reasonable excuse" for the offence occurring. This is not an easy defence to make out, and many offenders over-estimate the reasonableness of their specific excuse. Where Parliament has made an offence one of strict liability, then they had a purpose in doing so. Therefore, a reasonable-excuse defence must be more than the normal run of excuses and must be something significant that truly moves beyond what Parliament imagined when it made the offence one of strict liability.

The main mechanism by which the harshness of a strict liability offence is ameliorated in practice is the discretion that an individual prosecutor has. So a LA policy should have a degree of flexibility within it to decide not to proceed with a prosecution or civil penalty in certain cases. This might be where the landlord is vulnerable or where they have been the victim of dishonesty at the hands of their tenants.

It is also important to guard against abusive or unreasonable prosecutions and penalties. The magistrates have a right to decline to hear cases and to stay them permanently if they consider the prosecution to be unreasonable. Section 127 of the Magistrates' Courts Act 1980 specifies that a prosecution must be commenced within six months of the offence being committed. A scenario in which abuse frequently occurs relates to this time limit. This occurs where a LA lays information of their belief that an offence has been committed before the magistrates, thus commencing the prosecution, just before the end of the six-month time period, but then sits on the papers and does not serve them on the defendant. This is done to allow for further time to investigate the possible offence and keep options open. However, it is also abusive, and the magistrates will not permit the prosecution to continue if they become aware of it (*R v Brentford Justices ex parte Wong*). However, it is worth remembering that a failure to have an HMO licence, for example, is an ongoing offence, and so the six-month limitation period does not start until the offence ceases to be committed.

The final problem that LAs should guard against is where they have encouraged or induced the commission of the offence. So, for example, if a LA officer was to incorrectly assure a landlord that his property did not require licensing, then that officer would be likely to be seen as having induced the commission of the offence. Therefore, an attempt to prosecute the offence would not be reasonable (*Wandsworth LBC v Sparling*).

All of the above issues should be embraced within any prosecution policy.

Prosecution or civil penalty

A LA should also have a clear policy approach which covers when they will prosecute and when they will issue a civil penalty. This policy will need to consider the appropriateness of enforcement action in the first place and should take into account all of the issues raised above as well as considering the evidence and public interest in line with the Crown Prosecution Service Code for Crown Prosecutors.

The very different criteria for prosecution and civil penalty must also be considered. There is an unfortunate tendency among some officers to believe that a civil penalty is an easier version of a prosecution. This is not the case. The test for commencing a prosecution is a reasonable belief that a prosecution will succeed. The test for issuing a civil penalty is somewhat higher, as the LA must believe beyond reasonable doubt that the offence has been committed. This is a far higher standard, and it is important that LA officers recognise that this means that they must have cast-iron proof before a civil penalty can be issued.

The other key point to consider is the very different further enforcement options that flow from the different mechanisms. Banning orders can only be obtained from a prosecution, not from a civil penalty. So if the aim is to remove the very worst landlords from the sector then a prosecution is the most appropriate course. If there is a desire to list the landlord or agent on the rogue landlord database then that requires two civil penalties within a relatively short time period but can be done after just one prosecution. Therefore, it is again beneficial in some cases to secure a prosecution in order to obtain a better longer-term enforcement position.

Post-prosecution publicity and the databases

Naturally after a prosecution or civil penalty there is a tendency to wish to obtain a degree of publicity. This is a natural and proper part of the regulatory enforcement process, as one of the legitimate objectives of enforcement is to encourage observance of the law by others who might be inclined to break it. Publicity naturally serves that objective by ensuring that a prosecution or penalty comes to the attention of those persons. However, a degree of balance is necessary, as publicity also serves as a further punishment, and that may not be warranted in all cases. A LA should have a policy as to what cases it chooses to publicise and by what means. In doing so it should bear in mind the possibility of abuse and threats against those who have already been punished and whether they or their family members are vulnerable to such threats. It is also important to bear in mind that the UK courts have ruled that there is a right of individuals to not have their crimes follow them around on the internet indefinitely (*NT1 and NT2 v Google LLC*). Therefore any policy which includes publicity on a LA website should include a process which allows for deletion after an appropriate time. Such a decision should be based on the seriousness of the offence, the level of penalty, whether there has been rehabilitation and whether there has been further offending. It is also important to consider the possibility of appeal in any publicity which should therefore be delayed until after all rights of appeal have been exhausted. A similar consideration should apply to television programmes or other publicity that attends an investigation. Television and news programmes will wish to broadcast their footage, but it should be borne in mind that the fact of an investigation does not denote guilt, and a prosecution may be jeopardised if publicity prejudices a fair trial.

As well as publicity in the traditional sense consideration should be given to using the various databases. For LAs in London the London mayor has a database of penalties and prosecutions. This has a public-facing element so the above considerations on publicity will be relevant to this. There should also be a consideration as to the appropriateness of adding a landlord or agent to the rogue landlord's database or banning them and adding them to the database in appropriate cases, and this should be linked to the overall aims and objectives of the LA's policy structure.

Conclusions

LAs have a duty to be dispassionate and fair enforcers of the law who act in a clear and transparent manner. This does not mean that they cannot have a degree of discretion and make pragmatic decisions in each case. But that decision-making should sit within a framework set out by the LA's policies.

References and legislation

Avon Ground Rents Ltd v Child [2018] UKUT 204 (LC). Online. Available: www.bailii. org/uk/cases/UKUT/LC/2018/204.html (accessed 19 October 2021).

Housing Act 2004. Online. Available: www.legislation.gov.uk/ukpga/2004/34/contents (accessed 4 November 2021).

Housing and Planning Act 2016. Online. Available: www.legislation.gov.uk/ukpga/2016/22/contents (accessed 4 November 2021).

London Borough of Islington v The Unite Group Plc [2013] EWHC 508. Online. Available: www.bailii.org/ew/cases/EWHC/Admin/2013/508.html (accessed 28 October 2021).

Magistrates Courts Act 1980. Online. Available: www.legislation.gov.uk/ukpga/1980/43/contents (accessed 4 November 2021).

Mohamed & Anor, R (on the application of) v London Borough of Waltham Forest [2020] EWHC 1083 (Admin) Online. Available: www.bailii.org/ew/cases/EWHC/Admin/2020/1083.html (accessed 29 October 2021).

NT1 and NT v Google LLC & Others [2018] EWHC 799 (QB) Online. Available: www.judiciary.uk/wp-content/uploads/2018/04/nt1-Nnt2-v-google-2018-Eewhc-799-QB.pdf (accessed 4 November 2021).

Protection From Eviction Act 1977. Online. Available: www.legislation.gov.uk/ukpga/1977/43 (accessed 4 November 2021).

R v Brentford Justices ex parte Wong (1980). 73 Cr. App. R. 67.

Regulatory Reform (Fire Safety) Order 2005. Online. Available: www.legislation.gov.uk/uksi/2005/1541/contents (accessed 4 November 2021).

Wandsworth LBC v Sparling (1988). 20 HLR 169.

6 The shadow private rented sector examined

Ben Reeve-Lewis

Introduction

Safer Renting is a housing advocacy service, an organisation committed to strengthening rights and access to justice for tenants exploited by rogue (or criminal) landlords so that private renting is safe for all. Safer Renting aims to:

- Prevent homelessness, particularly by intervening in illegal evictions
- Support private renters to negotiate better conditions in their homes
- Enable private renters to leave a criminal landlord on their own terms, with compensation whenever possible
- Inform government policy and best practice by analysing and communicating how criminal landlords are exploiting the London housing crisis at the expense of tenants.

(for more information see: https://ch1889.org/safer-renting)

LAs traditionally employed tenancy relations officers (TROs) to help support tenants threatened with eviction and to help advocate for better housing conditions. More recently fewer LAs provide TROs, and in the absence of this, Safer Renting now offers these services on a contractual basis to many LAs so that PRS tenants are offered some support.

This chapter will examine the part of the private rented sector exploited by criminal landlords and adversely affecting multiple tenants' health, safety and security. It draws from both literature and practice experience at the front line and in particular the findings from the recent report on the so-called *shadow private rented sector* (Spencer et al., 2020).

The changing face of the private rented sector: challenges of enforcement

Figures reported in the census of 2011 showed 4.2 million households living in the PRS, amounting to around 16.3% of housing tenure in the UK,

DOI: 10.1201/9781003246534-7

representing an increase of 106% since 2001. By the years 2011–2012 the PRS had grown to overtake the social rented sector (Chan and Thompson, undated). Numerous factors account for this: the selling of social housing without replacement, the rise in the buy-to-let mortgage market, a serious shortage of housing supply in different parts of the UK, particularly cities.

In a market driven by supply and demand, acute housing need has also pushed up rents in many areas, particularly London. The prospect of high rental yields is proving attractive to many property owners looking for investment opportunities and, as with any fast-growing and lucrative market, has also attracted owners and agents with questionable motives and outright criminal intent.

At exactly the same time that these opportunities to exploit the sector have been growing, staffing levels among local authority enforcement teams, responsible for policing the private rental sector, have experienced cuts. To illustrate this, Battersby (2018) reported that:

> The average number of Environmental health officers available to inspect and enforce in respect of private rented accommodation was 2.46 in London and 2.2 for 10,000 properties . . . It would be difficult for these staffing levels to cope with complaints alone never mind take the initiative and seek out the criminal landlords.

These figures do not take into account the reduction in LA officers carrying out tenancy relations officer (TRO) functions in respect of dealing with incidents of harassment and illegal eviction. This function is not a statutory duty on a LA, and as a result of austerity-driven cuts, aimed in no small part on reducing staffing budgets, many LAs have no genuinely operational TRO function at all. In these districts illegally evicted households will find themselves drawing on homelessness assistance because there are no officers skilled or experienced in protecting tenants from harassment or gaining re-entry in cases of illegal eviction.

The twin attractions of a lucrative market for those prepared to exploit the opportunities without regard to law or safety and the lowered risk of detection and prosecution are proving increasingly compelling for many of the worst operators.

Why use the term "criminal" as opposed to "rogue"?

It is important for this author to make clear that a distinction should be made between the accepted term "rogue landlord", originally coined by Shelter in their 2011 campaign, and "criminal activities" engaged in by a wider group of individuals and companies. The term "rogue landlord" has always been ill defined and to a large degree unhelpful (see, for example, its use in MHCLG, 2019). Many breaches of landlord and tenant law are criminal offences, and it is accepted by all that due to the complexity of landlord and tenant legislation, landlords and agents break laws they do not even know were there and will quickly address their position once properly advised.

In this chapter we use the term "criminal" because the intent behind the activities referred to is characterised by individuals who set out deliberately to operate and create business models that use evasion and exploitation of the law to maximise their income, irrespective of enforcement and licensing provisions or the safety and welfare of their renters.

Safer Renting's 2020 report, "Journeys in the Shadow Private Rented Sector" (Spencer et al., 2020), delved further into the profiling of criminal landlords and identified five distinctly nuanced groups:

- **Willfully ignorant** landlords who tended to have small portfolios and were letting with no intention of meeting their statutory obligations
- **Corner cutters** had larger portfolios and maximised their rental income through noncompliance, factoring penalties and fines into their business model.
- **Scammers** remained hidden and often used the internet to swindle tenants – and landlords – through securing and then stealing deposits or renting property that was immediately sublet or let on the short-let market.
- **Prolific offenders** showed a blatant disregard for the law, often acting unpleasantly and with impunity, and were confident about their ability to challenge any attempt at prosecution.
- **Letting linked to organised crime** in which letting might be associated with labour and sex trafficking and the use of rented property as cannabis farms.

These profiles more accurately reflect the day-to-day activities and motivations of the subject group as distinct from "rogue landlords", who – as already mentioned – often fall into criminal acts through ignorance.

In this respect our definition is more in keeping with what is sometimes known as "Rachmanism", except insofar as the current criminal market is a lot more varied in scope than that of the late 1950s, involving fraud as much as anything else.

Routine yet endemic practices

Whilst there are specific criminal activities engaged in, there is also a range of common practices used by the majority of criminal operators to evade identification and detection and thus avoid redress and sanctions as follows:

Identity confusion

Criminal operators will routinely use aliases and chains of different company responsibilities to confuse both renters and enforcement officers alike. Renters will often be completely ignorant of the identity of the property owner or the arrangement by which they came to be renting, which may or may not have been authorised by the owner. Landlords and agents use different combinations of their given name and will just as commonly only be known to their renters by one name, which may well be false.

Quite apart from the way this simple device dupes renters and can leave them without effective means to claim back their deposit, it can also prove very problematical to local authority enforcement officers and legal representatives of tenants in identifying parties to be recorded on statutory notices or be named in legal proceedings.

The author of this chapter has encountered more than one letting agency where each employee in the company gave different names to different tenants. One tenant may know an individual agent as Mark, whilst they were known to another tenant as John, so just identifying who the complainant had spoken to in the agency was difficult, if not impossible.

Who is the landlord?

It is a common misperception that a property owner will automatically be the landlord, but the legal standing of a landlord in law is far from simple and is the subject of much debate. As Solicitor Robin Stewart (2019) commented in a blog article:

> Litigators have been wrestling for centuries over issues such as the identities of parties to contracts and the circumstances where agents assume personal responsibility for liabilities under a contract they entered into on behalf of their principal. From a contract law point of view this is already fiendishly complicated even before getting into matters such as "tenancies by estoppel".

Many breaches of legislation that are prosecutable by the local authority or which allow for tenants to take civil redress rely on identifying the "landlord", but unfortunately where the identity of the landlord is unclear, redress is thwarted.

Confusing management responsibilities

Running alongside the obfuscation of identities and possible landlord status is the widespread practice of creating murky webs of management responsibility, one agent to source a tenant, one to deal with repairs, one to accept the rent etc. Such companies will also change over time to create yet more confusion. Whilst outsourcing different management functions is not illegal it does again provide a further smokescreen for criminals to hide behind.

In cases involving numerous tenants in houses in multiple occupation, different agencies can be involved with each individual, leaving anyone trying to provide assistance for a group of renters having to deal with several companies at once. When this is added to the above reported practices of using aliases and confused landlord status, the tenant or enforcement officer is presented with a mountain to climb just to clarify who is really who.

Missing documentation

Landlords and agents operating in the shadow PRS will commonly fail or just refuse to provide written tenancy agreements or receipts for rent to reduce any paper trails that might lead local authority enforcement or HMRC officers to take an interest in their affairs.

Renters who find themselves faced with unemployment and needing to involve the landlord or agent in making a claim for benefits will often find themselves summarily evicted, because the continued success of a property run in this way relies on being completely under the radar of investigating authorities.

Criminal landlords and agents will commonly refuse to provide receipts for rent paid in cash. Tenants clearly know this is not right but are threatened with eviction if they complain.

Deposit protection and Rent Repayment Orders

There are statutory duties incumbent on a landlord to protect certain tenants' deposits which are also routinely flouted, but the problems outlined above with clearly identifying the person responsible often deny tenants the redress afforded to recover their deposit or seek a penalty for non-compliance. As with the failure to provide receipts for rent, how does a renter claim back money they paid when they were not given a receipt to show they paid it or they cannot even identify the person they paid it to?

Similarly the relatively new development of Rent Repayment Orders (see also Chapter 7), allowing for a renter to claim back 12 months' worth of rent in specific circumstances, is often thwarted by simply not being able to identify the correct respondent for the action, who must be "the landlord" because so many dubious names and chains of management companies are utilised and, where they have no receipts for rent, can't satisfy the test of claiming back "rent paid".

The scourge of rent-to-rent scamming

Special mention must be made of this particular social ill. In the past 8 years or so this relatively new phenomenon is by far becoming the most troubling example of criminal activity in the PRS because of the amount of people affected by a single incident. Conventional rent-to-rent business models are simply where an owner leases their property to a management company or local authority for a given period to use for their own ends, largely removing any liabilities or responsibilities for the period of the lease. Such an arrangement will usually afford the leaseholder landlord status.

The dubious version of this comes in different forms. The leaseholders could let a property to a single person, who sub-lets the property without knowledge or approval of the owner, leaseholder or agent, by bringing in a significantly larger number of sub-tenants, in the process often creating an overcrowded, unlicensed house in multiple occupation. In the alternative it is not at all uncommon for the leaseholders or agent themselves to let to more people than would be allowed, sometimes even sub-dividing rooms in a property they do not own with stud walling to create smaller rental units and maximise rental income.

In yet another version of this model, owner and managing agent collude to increase rental yield, and when a property is discovered by authorities, both parties deny responsibility and provide confusing and contradictory information about the true nature of the management and leasing arrangements, making it difficult for renter and local authority alike to identify the relevant parties liable for legal action.

Discovery of unauthorised and often unlicensed sub-lets regularly leads to serious harassment and illegal eviction by parties engaged in this rent-to-rent model as a means of removing evidence and discouraging cooperation by occupants.

Harassment and illegal eviction

These are criminal offences under the Protection from Eviction Act 1977 and are used routinely by landlords and agents who have no intention of seeking possession through the courts. However, they are also just as routinely used by criminals renting out overcrowded and unlicensed properties who have been identified by local authority enforcement teams, such as environmental health, planning or licensing, with the aim of dispersing potential witnesses and eradicating evidence that might further compromise their activities.

Evicted households owed a homelessness duty at least have a certain safety net, but those owed no statutory duty will find themselves street homeless or sofa surfing as a result of the eviction. One of the effective options for seeking re-instatement is for an injunction to be applied for and served on the perpetrator, but to be done effectively, the services of a solicitor need be employed, and despite legal aid being available for re-instatement after an illegal eviction, it is rare to find a legal aid practitioner who has capacity to take on a case in emergency circumstances.

Whilst an evicted renter could conduct their own litigation they have no costs protection and can be hampered by having English as a second language or no spoken English at all.

Affected renters

By and large but not exclusively, these criminals choose to operate at the bottom end of the rental market, where the demographic is marked by low incomes or incomplete immigration status.

Safer Renting set out the following in their report (Spencer et al., 2020), drawing on data from 259 cases of households experiencing harassment, illegal eviction or exposure to fraudulent activities by landlords and agents, pursued between January 2018 and March 2020:

- 40 per cent of cases were BAME
- 42 per cent were white "other" (within this group, two-thirds were Eastern European)
- 17 per cent white British and 1 per cent "other"
- 47 per cent were single-person households in house shares
- 27 per cent were families with dependent children (half of these were lone parents)
- 11 per cent were single people
- 8 per cent were couples in non-sharing accommodation
- 4 per cent were adult multi-generational households
- 2 per cent were "other"

Low incomes were the most significant drivers, meaning the renters did not have the money to rent properties from more upmarket landlords and agents.

The growth in HMOs and the concomitant growth in rent-to-rent scamming detailed above has been in no small part enabled by high rents. The working poor, those in the gig economy can often afford no more than a few hundred pounds in rent. In many inner-city areas, particularly London, rent for a one-bedroom flat can top £1,400 per month, putting self-contained living outside of the person's reach. Conversely, turning a three-bed family home into a seven-bed HMO at £400 per room will easily double the rental income for the landlord or agent, so the criminals set out to provide accommodation for the poorest and those with little choice about where they live or the conditions they live in.

Some examples of recent cases

In March 2021 the London Borough of Waltham Forest successfully took legal action against landlords Lahrie and Shahira Mohammed, whose property empire stretched, by the landlord's own estimation, to 600 properties, many of which are houses in multiple occupation (Wolek, 2020).

The London Boroughs of Westminster and Haringey combined to take successful legal action against portfolio landlord Ms Katia Goresmsandu following 60 separate and consecutive housing breaches to protect her tenants living across two boroughs, which resulted in a 10-year-long Criminal Behaviour Order being imposed against her (London Property Licensing, 2017).

In June 2021 the London Borough of Camden was forced to obtain an Anti-Social Behaviour injunction against Mohammed Ali Abbas Rasool, placing exclusion zones with powers of arrest around properties he ran where he repeatedly harassed and illegally evicted his tenants. Assisting the council with the court case the Metropolitan Police revealed that they had received similar complaints against Mr Rasool in other London boroughs as well (Camden Newsroom, 2021).

The above represents just three landlords adversely affecting the lives of hundreds of renters at any one time.

Conclusions

Rising rent levels and the shortage of affordable housing, coupled with low staffing levels for policing the PRS and unwieldy legislation, have created a perfect breeding ground for criminals to operate in what they see as a lucrative and comparatively risk-free market. Whilst criminal activity is not new in the PRS it has grown exponentially since the financial problems of the past 12 years began, and the nature of it has changed; individuals taking their chances have been supplanted by systematic and organised criminal enterprises.

Housing enforcement officers are having to develop fraud investigation skills of their own that were not generally needed until a decade ago. It has been difficult to estimate an accurate number of criminal landlords currently operating and the effect their operations continue to have on the lives of tenants, as well as the length of time taken for LAs and organisations such as Safer Renting to investigate and intervene in such complex cases. This author contests that to count the number of criminal landlords is to reverse the real test. What is more appropriate is counting the number of their victims, present, previous and future and the amount of chaos a willful landlord or agent can cause a community.

Prosecution rates and other penalties, while commendable and encouraging, do not provide full insight into the size of the problem for all of the reasons set out in this chapter. The numerous fraudulent activities practised by these landlords and agents protect so many against sanctions that might provide reliable data. To combat this it is essential that government recognises the financial commitment needed to properly fund enforcement, and more local authorities need to adopt multi-disciplinary methods of targeting criminal landlord activity. Silo working is still, unfortunately, far too widespread.

Drafters of legislation need to be more awake to the criminal practices used in the private rented sector. The judiciary, when hearing applications for sanctions and appeals, similarly need to support local authorities where the evidence is compelling that smokescreens are being employed by criminal operators.

This combination of weaknesses in the system, coupled with the games that committed criminals play, will all too often leave those charged with policing the PRS always one step behind.

References

Battersby, S. (2018). *Private rented sector inspections and local housing authority staffing.* Online. Available: Available www.sabattersby.co.uk/documents/Final_Staffing_Report_Master.pdf (accessed 29 September 2021).

Camden Newsroom. (2021). Camden secures first injunction against landlord to stop illegal eviction of tenants. *Camden Newsroom Blog.* Online. Available: https://news.camden.gov.uk/camden-secures-first-injunction-against-landlord-to-stop-illegal-eviction-of-tenants/ (accessed 23 November 2021).

Chan, D. and Thompson, M. (undated). *Understanding the growth in private rented housing.* Core Cities Report. Online. Available: www.corecities.com/sites/default/files/field/attachment/UnderstandinGrowth%20in%20Private%20Rented%20Sector.pdf (accessed 1 October 2021).

London Property Licensing. (2017). *London landlord issued with 10 year criminal behaviour order.* London Property Licencing Blog. Online. Available: www.londonpropertylicensing.co.uk/london-landlord-issued-ten-year-criminal-behaviour-order (accessed 24 October 2021).

MHCLG. (2019). *Rogue landlord enforcement: Guidance for local authorities.* London: HMSO.

Spencer, R., Reeve-Lewis, B., Rugg, J., and Barata, E. (2020). *Journeys in the Shadow Private rented sector*. Cambridge House and University of York. Online. Available: https://thinkhouse.org.uk/site/assets/files/2210/ch0920.pdf (accessed 10 October 2021).

Stewart, R. (2019). Who can a rent repayment order be made against? Who is the landlord? *Anthony Gold Blog*. Online. Available: Available https://anthonygold.co.uk/latest/blog/who-can-a-rent-repayment-order-be-made-against-who-is-the-landlord/ (accessed 14 October 2021).

Wolek, A. (2020). *Mohamed & Lahrie v London Borough of Waltham Forest and Secretary of State for Housing, Communities and Local Government, Sharpe Pritchard Blog*. Online. Available: www.sharpepritchard.co.uk/latest-news/mohamed-lahrie-v-london-borough-of-waltham-forest-and-secretary-of-state-for-housing-communities-and-local-government/ (accessed 24 October 2021).

7   Making more effective intervention choices

David Beach

Introduction

This chapter explores some of the choices available to LAs in administering their private-sector housing strategies. It includes an overview of civil (financial) penalties and the burden of proof necessary in delivering this, an overview of options available and how they work in practice. In this chapter we look at Civil Penalties, Rent Repayment Orders and Management Orders.

Civil (financial) penalties

LAs need to have strategies in place to effectively deliver private-sector housing regulation. For the purposes of this chapter, enforcement policy is key and includes both criminal and civil remedy. The Housing and Planning Act 2016 provided a suite of options including some around to support LAs in delivering their statutory and discretionary functions, including provision for civil penalties. It must be remembered that in some cases, criminal remedy should be the preferred option since it may lead to a more appropriate outcome in relation to a particular landlord, such as being included on the Rogue Landlord database, which a civil remedy alone may preclude. Other enforcement tools include Interim (and Final) Management Orders and Rent Repayment Orders as well as wider powers around anti-social behaviour.

The Housing and Planning Act 2016 introduced civil penalties as one of a number of sanctions directed toward rogue landlords. These were available as an alternative to prosecution proceedings for certain specified offences as follows:

- Failure to comply with an Improvement Notice
- All licencing offences (Part 2 and Part 3)
- Failure to comply with Overcrowding Notice

DOI: 10.1201/9781003246534-8

- Failure to comply with HMO Management Regulations
- Breach of Banning Order

Despite being a civil penalty, the same criminal burden of proof is required to show the relevant offence committed. The government issued guidance (DCLG, 2018) for LAs to use in exercising their powers to impose financial penalties. The initial 2017 guidance made it clear that ministers expected this power to be used robustly to clamp down on rogue landlords. The guidance also referenced the fact that the original maximum level of civil penalties had been increased to a maximum of £30,000 in order to provide a significant deterrent to landlords who flouted the law.

Local authorities must have regard to the guidance. It provides for a maximum of £30,000 per offence, such as each breach of The Management of Houses in Multiple Occupation (England) Regulations 2006 or The Licensing and Management of Houses in Multiple Occupation (Additional Provisions) (England) Regulations 2007. The Guidance sets out factors that should be taken into account when determining penalty levels, and there is no minimum penalty set. Separate penalties can be imposed on landlord and agent for the same offence.

There must be proof of offence 'beyond reasonable doubt', and the LA needs to be satisfied that the case meets an evidential test and a public interest test. Prosecution may be most appropriate option where offences are particularly serious but civil penalties can also be used where offences are considered serious. The guidance indicates that LAs should use powers to ascertain landlords' assets, not just the rental income.

Factors to be considered in determining a civil penalty should take into account:

- Severity of offence
- Culpability and track record of landlord; history of failing to comply or should have known
- Harm to the tenant
- Punishment of offender
- Deter offender from re-offending
- Deter others from committing an offence
- Remove any benefit in offending; it should not be cheaper to offend.

LAs need to decide on penalties in each case. This has proven challenging, with different authorities taking different approaches. However, many authorities have decided on a 'matrix approach' to split the £0 to £30,000 range into smaller bands with aim of achieving consistency. There are two basic approaches. First, LAs might set bands according to the size of a landlord's business, nominate bands for each offence and consider aggravating/mitigating factors. Second, they might allocate bands to each offence and adjust the band to the type of landlord and consider aggravating/mitigating factors. Whatever decision is taken, it must adhere to the guidance and be clear and accountable.

Good governance is important, and elected members should have the final say. There is no right or wrong approach, but it is very important that the LA is consistent in its use of civil or criminal proceedings and why this is so. It is desirable that penalties between LAs are consistent, but this presents challenges, as each has its own strategic approach.

Enforcement policy and civil penalties

Enforcement policies set out what residents, businesses and consumers can expect from enforcement officers, and it is important that enforcement is both consistent and transparent. Generally speaking, LA private-sector housing strategies are likely to prioritise protection of public health, reduce anti-social behaviour and safeguard housing standards through ensuring compliance with relevant legislation.

Notice of intent

The procedure for imposing a civil penalty is set out in the Housing Act 2004 Schedule 13a. Before imposing a penalty, the authority must give Notice of Intent (NOI) setting out the reason for the proposed penalty and the amount. This NOI must be within 6 months of offence, and a period of 28 days starting the day after the Notice of Intent is issued must be given to allow representations to be made. Following the representation period, the LA may decide to impose a 'Final Notice' imposing the penalty. There is a right of appeal to the Residential Property Tribunal (RPT).

There is no prescribed format of NOI, and so it is of little surprise that penalties have been challenged. The following cases are worthy of consideration here. In *Waltham Forest LBC v Younis (2019) UKUT 0362 (LC)* it was decided that brief reasons are acceptable, and reasons can be incorporated within separate documents such as attached witness statements. Furthermore, even if reasons are defective it will not usually justify invalidating the notice, particularly where there has been no prejudice to the appellant. In *Bharadwaj v Havering* the civil penalty was quashed as the LA had not given 28 + 1 day for representations.

The representation period is important and allows the recipient to make a case before any decision to impose the penalty via a 'Final Notice'. It is important for proper governance that any representations received are considered and answered. In some ways this is equivalent to initial plea hearings for cases following prosecution route, with grounds of defence identified. When it works well, this enables LAs to form a view not to pursue cases that are not merited and can identify key points of contention for cases that do proceed to Final Notice. Penalties can also be varied, but a disadvantage is that it can allow landlord to 'change horses'.

Financial penalties and appeals

Since penalty levels can be high there are a large number of appeals. The RPT can issue Directions, with dates for hearing bundles to be exchanged. Directions are fairly standard, and the LA needs to provide evidence to prove the offence was committed, that they have adopted a policy and how they have applied this to calculate the penalty. The Applicant provides expanded appeal grounds. Both sides give evidence and are cross-examined, and a decision follows within a target time of 6 weeks to uphold, vary or quash the penalty.

The appeal is a re-hearing of the LA's decision to impose a civil penalty. It may also have regard to matters of which the LA was unaware when the decision to impose a civil penalty was made. It can dismiss an appeal if satisfied that the appeal is frivolous, vexatious or an abuse of process or has no reasonable prospect of success. Appeals must be within 28 days, but there is discretion to extend the time to appeal, using its case management powers; see, for example, *Pearson v Bradford MDC (2019)*. However, Haziri v LB Havering (2019) held that the First-Tier Tribunal (FTT) should not consider the underlying merits of the appeal, and time should not be extended lightly.

The FTT can vary decisions. *Waltham Forest LBC v Marshall/Ustek* (2020) clarifies the correct approach to be taken by the FTT on appeals against penalties; this is the case of two penalties for failing to licence under a selective licencing scheme. In particular:

- the FTT may not entertain any challenge to an LA's enforcement policy but must usually enforce and apply it
- it must afford the LA's decision on the penalty 'special' and 'considerable' weight
- the appellant bears the burden of persuading the FTT to depart from the policy or the LA's decision.

There is no requirement for penalties to include discounts, although some LA do offer this with the intention of encouraging compliance and early payment. There needs to be an end-to-end process in place to secure payment of civil penalties such as a debt-collection service and agreement of suitable payment plans.

Are civil penalties a useful tool?

Civil remedies can help ensure a more consistent approach to issuing penalties. Monies collected are ring-fenced for housing enforcement. However, civil penalties carry less weight than a criminal conviction, and it is time-consuming to defend appeals. It is important to remember that the FTT is generally a 'no-costs' jurisdiction, and the cost to the LA in defending any appeal brought by a landlord against a civil penalty will generally not be recoverable.

Rent Repayment Orders

A Rent Repayment Order (RRO) is an order made by the FTT requiring a landlord to repay a specified amount of rent. An RRO may be applied for in one of several specific situations, and the maximum amount of rent that can be recovered is capped at 12 months. Repaid rent can be paid to the tenant or the LA. If the tenant paid their rent themselves, then the rent must be repaid to the tenant. If rent was paid through HB or through the housing element of Universal Credit, then the rent must be repaid to the LA.

RROs were introduced in Housing Act 2004 to provide an extra sanction where landlords had not obtained a property licence and were significantly extended by the Housing and Planning Act 2016 to include:

- Failure to comply with an Improvement Notice or Prohibition Order under Part 1 of the Housing Act 2004 (HHSRS)
- Breach of a Banning Order
- Using violence to secure entry to a property under section 6 of the Criminal Law Act 1977
- Illegal eviction or harassment of the occupiers of a property under section 1 of the Protection from Eviction Act 1977.

RRO process

The government introduced guidance in 2017 which LAs must follow (DCLG, 2017). RROs normally follow a conviction or a civil penalty that is not challenged, but an RRO may be applied for before a conviction. If there is no conviction, FTT will need to be satisfied that offence committed to criminal burden of proof. RROs do not automatically follow from a conviction, and a separate application must be made. Recent case law (*Rakusen v Jepson & Ors, Safer Renting Intervenor* (2021) EWCA Civ 1150) has confirmed that RRO applications should be made against the immediate landlord where there is both an immediate and superior landlord.

Before applying for a RRO, the LA must give notice of intended proceedings that:

- Informs the landlord of the intention to apply for RRO and why
- State the amount that the council is seeking to recover
- Invite the landlord to make representations within a period which must be at least 28 days.

The NOI must be within 12 months of offence date, and all representation must be considered before application for an RRO. A tenant not required to give NOI can apply to FTT. Generally, each side bears its own costs unless there is unreasonable behaviour.

Payments under RROs

The starting point for assessing the level of RROs is the full rent. The judgment in *Vadamalayan v Stewart and others* (2020) UKUT 0183 (LC) changed the ground rules on the approach to such matters that had been established in *Parker v Waller & Ors* (2012). The judgment in *Vadamalayan v Stewart* reflected the changed landscape around RROs following the introduction of the Housing and Planning Act 2016. Subject only to relief for exceptional circumstances, the order must be for the maximum amount which the FTT has power to award, even where the landlord has also been convicted and fined or received a CP.

Powers to take over the management of residential accommodation

The Housing Act 2004 Part 4 includes provisions relating to the making of Interim (IMOs) and Final Management Orders (FMOs). The act details the circumstances in which an LA is under a duty to take action using IMO or FMO powers and others when it has a power to do so.

LAs must make an IMO if either:

- the property should be licenced but is not and there is no reasonable prospect of being licensed in the near future; or
- the health and safety condition is met.

The LA must also make an IMO on revocation of a property licence if either:

- On the revocation coming into force, there would be no reasonable prospect of being licensed in the near future; or
- On the revocation coming into force health and safety conditions are met.

When such a Management Order comes into force, the LA must:

- Take immediate steps to protect the health, safety and welfare of those living in the property and those living in or owning properties in the vicinity
- Take steps to ensure the property is properly managed
- Ensure the property is insured against destruction or damage.

The Management Order gives the LA the following rights over the property:

- Possession of the property (subject to the rights of any existing occupiers)
- All the rights which a landlord has, including collection of rent
- The power to spend the money received through rent to carry out its duties to manage the property
- The power to grant new tenancies and licences, with the consent of the landlord.

There are, however, many challenges to LAs exercising IMO/FMO duties and powers. These include financial risk, the LA taking on the management responsibilities, including regular inspections and necessary works, tenancy management and setting up rent accounts.

The London Borough of Waltham Forest

The London Borough of Waltham Forest has been recognised for its proactive work in IMOs and FMOs including establishing an in-house lettings team (Lettings Waltham Forest – LWF), extending its social housing services, capitalising on its extensive experience of tenancy management including setting up rent accounts and debt management and working closely with the PRS housing team.

The use of LWF to manage properties that are subject to IMOs or FMOs aligns with wider licensing objectives to help ensure that the local PRS is effectively managed and that the standard of properties is improved through effective management and maintenance. It provides an increase in housing provision at a reasonable market rent. There has been a reduction in ASB in the community as a result of sustainable tenancy management and a corresponding increase in tenant satisfaction.

The PRS team has the client role, which includes identifying properties, drawing up IMO paperwork, defending any appeals, obtaining warrants as necessary, attending site to confirm works and draw up schedules, notifying Land Charges etc. The LWF role includes signing up tenants, arranging lock changes, managing tenancies, maintaining rent accounts and carrying out regular inspections.

Waltham Forest has used IMOs to achieve the following outcomes:

- The licensing of an address whose owner was not resident in the UK following threats to make an IMO after other warning letters had been ignored over an extended period
- Taking control of a flat from a landlord who was not performing any active management role, having previously failed to comply with an Improvement Notice served under Part 1 – bringing the flat under responsible management control.

Conclusion

Each LA will have its own PRS strategy utilising criminal and civil remedy. This chapter has focused on a range of options available and their practical and effective application.

References

DCLG. (2017). *Rent repayment orders under the Housing and Planning Act 2016: Statutory guidance for local housing authorities on the extension of rent repayment orders.* London: HMSO.

DCLG. (2018). *Civil penalties under the Housing and Planning Act 2016: Statutory guidance for local housing authorities on civil penalties.* London: HMSO.

Parker v Waller & Ors (2012). UKUT 301 https://www.casemine.com/judgement/uk/5a8ff82e60d03e7f57ebc3e3).

Rakusen v Jepson & Ors, Safer Renting Intervenor (2021). EWCA Civ 1150 https://www.bailii.org/ew/cases/EWCA/Civ/2021/1150.html.

Vadamalayan v Stewart and others (2020). UKUT 0183 (LC) https://www.casemine.com/judgement/uk/5f0fe3942c94e065efde0014.

8 The Housing Health and Safety Rating System (HHSRS)

A practitioner's perspective

Alex Donald

Introduction

At the time of writing, the Housing Health and Safety Rating System (HHSRS) has been enshrined in law for 17 years. We are waiting for the newly formed Department for Levelling Up, Housing and Communities (DLUHC) to finalise its consultation results and produce new guidance and training to make the system easier for EHPs, landlords and tenants to use and understand.

In light of the timing of this book, this chapter has been particularly challenging to write, since the HHSRS Review remains underway and we do not yet know its outcome. Therefore, this chapter explores and assesses this revolutionary risk-based system and draws together some of the critiques leading to the review, grouping them into three categories: (1) guidance, (2) direction and (3) formation.

Background to minimum standards versus HHSRS

Before we delve into HHSRS, it is worth noting that Scotland and Northern Ireland use minimum standards as a framework for assessing housing conditions. Since the Housing (Scotland) Act 1987, Scotland's EHPs have been using the Tolerable Standard (Shelter Scotland, 2021) with Northern Ireland's EHPs using the similar Housing Fitness Standard (DfC) under the Housing Order (Northern Ireland) 1992. Before 2006, England and Wales used the Fitness and Decent Homes Standard under Section 604 of the Housing Act 1985 (HoC Library, 2021), known as the statutory standard of fitness and a broadly similar type of assessment.

These systems are all a tick-box exercise resulting in a property receiving a pass or fail. If a property doesn't fulfil one or more of a set list of standards (e.g. be structurally stable), it is deemed to be unfit for human habitation. Northern Ireland is considering adopting the HHSRS, and a bill will be brought forward in due course (HoC Library, 2021).

DOI: 10.1201/9781003246534-9

In contrast, the HHSRS is the approach now used in England, Wales and the United States (where it is called the Healthy Home Rating System or HHRS). This method evaluates the potential risks to the physical and mental health, safety and social well-being of occupants or visitors from any deficiencies identified in dwellings. The original 2006 Operating Guidance (referred to in this article as ODPM (2006)) explicitly calls cost "irrelevant to the assessment" (ODPM, 2006: 9).

Underpinned by scientific and statistical data of which "it is the responsibility of professionals using the HHSRS to keep up to date on current evidence", the central principle is that "any residential premises should provide a safe and healthy environment for any potential occupier or visitor" (ODPM, 2006: 7–8). This ranges from Victorian houses to modern flats, and the housing data is tailored accordingly.

Originally proposed in 1998 and enshrined in the Housing Act 2004, the HHSRS groups any number of deficiencies into an exhaustive list of 29 hazards. These hazards are themselves grouped into four different profiles – those that could cause physiological harm (e.g. Damp and Mould), psychological harm (e.g. Noise), infection (e.g. Domestic Hygiene, Pests and Refuse) and accidents (e.g. Falling on Stairs, etc.).

Regardless of actual occupation, assessments must prioritise the effects on under-18s and over-65s (or the so-called younger and older vulnerable age groups, respectively). Vulnerability does not cover those with disabilities, and if the hazard data does not specifically mention age, the population is taken as a whole (ODPM, 2006).

Using the data, each hazard is assessed according to the likelihood of an incident in the 12 months following the assessment. The likelihood of a fall from an indoor staircase that lacks a lower guarding is 1/180, whereas that of a fall from a similarly deficient external staircase is 1/18. The second scenario is scored much higher because the external staircase is exposed to the elements and would be particularly slippery in winter, especially for an older person, and the impact onto a concrete surface would be more serious than that onto interior flooring.

The assessment also scores the harm outcome of a potential incident ranging from moderate (e.g. bruising) to extreme (e.g. death). The example given by ODPM (2006) is that of a loose window with a low internal sill on the ground floor compared with the same on the second floor. A child falling from the ground floor window is 99% likely to suffer bruising, whereas from the second floor they are 80% likely to suffer serious fractures and even 10% paralysis or death. Despite the same tragic likelihood, the second scenario is scored higher due to the higher chance of a more fatal outcome.

A set of Worked Examples (WEs) presenting the statistics on likelihood and harm outcomes is provided for each hazard so that EHPs can make as accurate an assessment as possible.

We now turn to criticisms in the form of Guidance, Direction and Formation around HHSRS.

Guidance

Some of the critiques of HHSRS can be attributed to a lack of information, understanding and training, which has led to frustrations among stakeholders.

Complexity and resources

One such criticism is that HHSRS is too complex and subjective and as a result leads to inconsistent and contradictory interpretations by different and invariably under-resourced EHPs across the country, even from within the same LA (HoC Library, 2021).

In a survey of stakeholders, 53% reported seeing hazards not adequately addressed by HHSRS. Among the 170 responses, there was a highlighted need to update the guidance for certain specific hazards (CIEH, 2017).

The low number of and lack of experienced EHPs (CAG Consultants, 2018) has only added to the perceived complexity of the system. With HHSRS being so subjective, EHPs have struggled with being able to clearly rate some hazards known to be higher risk (Universities of Kent and Bristol, 2017). Many have argued that a lack of resources has led to reactive cultures in which enforcement is either too much or too little. In some cases, EHPs are (very wrongly) taking informal action over Category 1 hazards (Battersby, 2011), and yet in others they are jumping straight to enforcement without giving the landlord any warning (CAG Consultants, 2018). Both have led to appeals to the First-Tier Tribunal (FTT).

Speaking of the court system (see Chapter 5), it appears that confusion has hindered EHPs from carrying out prosecutions, as 80% of LAs state they have not made a single one (Battersby, 2011). Could this be because under-resourced EHPs pull their punches for fear of long-winded appeals that eat into their time? It also does not help when the FTT gives inconsistent and unhelpful rulings (see 'Affordability').

Outdatedness

The data supporting HHSRS has not been regularly updated and yet continues to form the basis of assessment. Despite the onus being on practitioners "to keep up to date on current evidence" (ODPM, 2006: 7), in practice this has proved to be unrealistic. Yet while statistics may be out there, updated guidance has rarely been issued by DLUHC or its previous incarnations.

Much of the available data on housing stock for HHSRS comes from the 1990s, and this is particularly unhelpful given the huge advances in domestic energy efficiency. Even in recent years the situation out there has improved immensely: only 13% of homes in the private rented sector were in EPC Bands A–C in 2009. In 2019, this almost doubled to 23% (MHCLG, 2020a). A 2018 report on fuel poverty in the private rented sector has put the unwilling freeze on new data for EHPs down to a lack of political will in tackling fuel poverty (CAG Consultants, 2018).

This could equally be applied to housing standards more generally: the WEs used by EHPs to assist in their HHSRS assessments have on the whole remained the same since their initial release in 2006 in conjunction with the system itself. With 9 in 10 respondents to the CIEH survey calling for a more up-to-date set, the only significant new WE came about in 2018 in response to the Grenfell fire disaster in the previous year, where 72 people perished. The new WE presented a scenario of new ACM cladding on a multi-storey tower block (HoC Library, 2019) (see 'External Cladding'). Calls to provide additional WEs, including in borderline cases, would be welcomed by EHPs. However, there are early indications that both Grenfell and the COVID-19 pandemic have increased the political will to tackle inefficient homes following the conservative government's ambition to reduce the UK's carbon emissions to net zero by 2050 (HM Government, 2021).

The Net Zero agenda (see 'Net Zero') is another factor that must be included in any new guidance. At the moment, case law and the general consensus by stakeholders are that gas heating should be prescribed to avoid excess cold in the home (CAG Consultants, 2018), contradicting Whitehall's stated aim to stop the installation or replacement of gas boilers by 2035 in favour of electric heat pumps (HM Government, 2021). This must be clarified.

Lack of awareness

So far we have covered the confusions felt by EHPs about HHSRS. However, there are other stakeholders to consider: what about HHSRS understanding (and compliance) amongst ordinary landlords and tenants?

If 100 landlords were asked to explain HHSRS, only 15 would have even heard about it (HoC Library, 2019). Out of 100 tenants, the number would be lower, as only 10 would have even the faintest clue on where responsibility lies for repairs (HoC Library, 2021). Many tenants are not aware of their rights (CAG Consultants, 2018), and one-third of landlords find it difficult to keep up with legislation (HoC Library, 2021).

HHSRS has been described as "unnecessarily complicated" for landlords and tenants. As previously argued, it is not the system itself but the delivery of the system that is causing confusion. Thankfully this has been recognised by the government in their review, calling HHSRS a "fundamentally sound" system (HoC Library, 2021).

These criticisms point to a 'sound' system failed by breaks in communication between central government, LAs, EHPs, landlords, tenants and the FTT. HHSRS is a complete system that requires regular updates clearly conveyed to practitioners and stakeholders. However, other criticisms indicate a clear deviation from its original intent.

Direction

The following critiques are grouped together because they follow case law rulings, legislation and other departmental guidance that have caused the 2006 system to misfire into completely different directions from its intended journey. While all these developments have undoubtedly been carried out with the best intentions and indeed produced positive outcomes for individuals, they have also unwittingly caused the risk-based system somewhat of an identity crisis.

Affordability

When Liverpudlian landlord Anwar Kassim installed expensive-to-run panel convector heaters and an electric towel rail as the sources of heat for his rented home, he may have just read section 1.18 of ODPM (2006) calling cost "irrelevant" to HHSRS. As previously outlined, the HHSRS should only be concerned with the health, safety and social well-being of occupants.

However, Liverpool City Council also cited the Guidance and made regard to the older vulnerable age group, arguing that they would disproportionately use the panel heaters less frequently to save costs. For EHPs in the council, this was a profoundly unfair situation that required enforcement against Mr Kassim following an assessment under the Excess Cold hazard.

Mr Kassim stood firm on his literalist position. Between 2011 and 2015, he and the council were involved in multiple tribunal appeal hearings until finally he was ordered to install gas central heating and reduce the running costs for the occupant. The Upper Tribunal decided that these comparative costs of heating systems were indeed a factor, as those in the older vulnerable age group are, in general, less well off than the population as a whole (CIEH, 2019).

As a result, *Liverpool City Council v Kassim* [2011] UKUT 169 (LC) (see Nearly Legal, 2012) has taken HHSRS down a completely different path from its original path and raises profound repercussions for issues very close to home. For example, cost and affordability are especially urgent given the climate emergency and the hotly contested debate on who will ultimately fund low-carbon alternatives to gas boilers (HM Government, 2021), as well as who is funding the replacement of flammable ACM cladding. Therefore clarity is sorely needed for EHPs on their powers in any new guidance.

Back to minimum standards

In England and Wales, housing enforcement legislation has produced a de facto hybrid system of risk assessment and minimum standards – a situation scorned by Ormandy and Battersby as "nonsense" and "a step back" (Ormandy and Battersby, 2020: 108–110). This has come in the form of instruments such as the Carbon Monoxide (England) Regulations 2015, which require rented properties to have one smoke alarm installed on every storey and one CO alarm in any room with a solid fuel–burning appliance. In addition, the Energy Efficiency (Private Rented Property) (England and Wales) Regulations 2015 has led to all tenancies requiring a minimum Band E EPC rating. Both of these regulations are in direct conflict with HHSRS. As explained at the beginning of this chapter, a set of 'tick-box' standards undermines the EHOs' judgement and ability to assess hazards on the basis of likelihood and outcome. It also ultimately creates confusion on top of an already under-resourced profession (see 'Complexity and resources').

Despite the scientific and accurate foundations of HHSRS, there have been many calls for England and Wales to rejoin Scotland and Northern Ireland in adopting a complete minimum standards system. Even a Parliamentary Select Committee has recommended HHSRS be scrapped and replaced with a more straightforward set of quality standards (HoC Library, 2021).

As explained in the previous criticisms from the Guidance, the main reasons for this can be traced to miscommunications of varying degrees. It is DLUHC's responsibility to ensure EHPs are given regular up-to-date guidance, housing data and WEs. It is also in LA's interests to be properly funded and resourced so that EHPs can do their jobs effectively.

Despite the Select Committee's recommendation, the government has opted to keep HHSRS and attempt to correct the disconnect between it and other existing legislation (MHCLG, 2019).

External cladding

We have touched on the Grenfell tower fire disaster in the context of how very few updates since 2006 have aided EHPs to carry out assessments. The new HHSRS Addendum and WE released in 2018 was a watershed moment for the risk-based system, as it deliberately shifted the 'predominant focus' from "within individual dwellings" to include the exterior of buildings (MHCLG, 2018). It is acknowledged that ODPM (2006) does not specifically cover cladding (HoC Library, 2019).

Again, few would criticise the government for at least attempting to address the appalling injustice where flammable cladding has cost the lives of 72 innocent flat-dwellers (and put many more throughout the country at risk).

As mentioned in 'Affordability', there is still an unanswered question related to replacement costs and why leaseholders are bearing the brunt. However, the main point of this section is to show how developments in case law and new legislation have negatively altered HHSRS from its original form, unlike the next section, where the original system itself comes under scrutiny.

Formation

These final critiques argue that although the HHSRS is by far the most effective system we have for assessing housing conditions, it could be argued that it was misformed, as it lacked within it certain aspects of equality and health which – if concocted today – would almost certainly be cemented at the very heart of the system. Thankfully, given the changes highlighted in 'Affordability', we know that it is not too late to right these wrongs.

Disability as vulnerability

All too often, our society overlooks the needs of people with disabilities, and HHSRS is no exception. As mentioned, we have seen how ODPM (2006) has been undermined by subsequent rulings of the FTT. Whereas in 2006, cost was explicitly "irrelevant to the assessment", today following updated case law, assessments can have regard to cost.

ODPM (2006) also states that age is the only vulnerability to which EHPs must have regard and explicitly rules out extending the concept "for other reasons". However, there is a strong argument to bring people with disabilities into this fold.

In its current form, HHSRS only allows EHPs to have regard to age in their assessments. It would not factor in people with disabilities – many of whom rely on lift access to an upper floor of a building. In the case of Grenfell tower, a report from the Universities of Kent and Bristol (2017) argues that such an assessment would have highlighted the risks posed to both occupiers and rescuers.

As with the vulnerable age groups, it is conceivable that the likelihood and harm outcomes would be adjusted regarding certain hazards, e.g. Falling on

Stairs (for wheelchair users) or Noise (for people on the autism spectrum). It is an injustice that this branch of equality is not embedded into the HHSRS.

Of course, this suggestion falls outside DLUHC's consultation remit (MHCLG, 2019). However, as has been demonstrated with *Kassim vs Liverpool City Council*, change can come in many forms, and for the sake of equality, this would be a welcome development.

Net zero

At the time of writing, world leaders are meeting in Glasgow at the 26th UN Climate Change Conference of the Parties (COP26) to discuss and agree on measures to reduce carbon emissions. The conservative government has released a report (HM Government, 2021) committing to making the transition to low-carbon buildings affordable and achievable for all. This includes installing 600,000 new heat pumps per year by 2028 and phasing out the installation of new and replacement gas boilers by 2035.

Unlike redefining 'vulnerability' to include disability, it is already established in the ODPM (2006) guidance that the HHSRS evaluates the harm to human health, and there is therefore no reason the effects from excessive carbon emissions from properties are not covered.

Creating tailored WEs to aid EHPs on the harm outcomes of gas boilers or the lack of electric vehicle chargers within the property boundary would not only increase joined-up thinking between government departments and stakeholders, but it would also encourage EHPs to be mindful of carbon emissions when carrying out their assessments.

Conclusion

This chapter has briefly explained HHSRS and introduced the various critiques from its supporters and detractors. It cannot be stressed enough its radical nature within the context of public health and housing enforcement, especially when considering 19th-century pioneers such as Edwin Chadwick, who could only dream of such a system after facing overwhelming political and social resistance for advocating basic domestic sanitary improvements.

However for HHSRS to work, it also requires constant reviews, which are here categorised and critiqued in three distinct groups in the hope of helping stakeholders to understand the challenges and influence the debate.

As we await DLUHC's new guidance and training following their extensive review, we can reflect on its journey and anticipate that that HHSRS can be refined and revived into an evolving and effective tool for making housing conditions fit for the 2020s and beyond.

References

Battersby, S.A. (2011). *Are private sector tenants being adequately protected: A study of the Housing Act 2004, Housing Health and Safety Rating System and Local Authority Interventions in England – A report by Dr Stephen Battersby for Alison Seabeck MP, Shadow Housing Minister & Karen Buck MP, Shadow Work and Pensions Minister.* London. Online. Available: www.sabattersby.co.uk/documents/HHSRS_Are%20tenants%20protected.pdf (accessed 13 August 2021).

CAG Consultants. (2018). *The warm arm of the law: Tackling fuel poverty in the private rented sector.* Online. Available: https://cagconsultants.co.uk/wp-content/uploads/2018/07/The-Warm-Arm-of-the-Law-report.pdf (accessed 20 September 2021).

Chartered Institute of Environmental Health. (2017). *HHSRS – 11 years on.* Online. Available: www.cieh.org/media/1166/hhsrs-11-years-on.pdf (accessed 5 October 2021).

Chartered Institute of Environmental Health. (2019). *CIEH excess cold enforcement guidance.* Online. Available: www.cieh.org/media/3762/cieh-excess-cold-enforcement-guidance.pdf (accessed 31 October 2021).

HM Government. (2021). *Net Zero Strategy: Build Back Greener.* Online. Available: https://assets.publishing.service.gov.uk/government/uploads/system/uploads/attachment_data/file/1026655/net-zero-strategy.pdf (accessed 31 October 2021).

House of Commons Library. (2019). *The housing health and safety rating system.* Online. Available: https://researchbriefings.files.parliament.uk/documents/SN01917/SN01917.pdf (accessed 5 October 2021).

House of Commons Library. (2021). *Housing conditions in the private rented sector (England).* Online. Available: https://researchbriefings.files.parliament.uk/documents/CBP-7328/CBP-7328.pdf (accessed 28 September 2021).

Ministry of Housing, Communities and Local Government. (2018). *Housing health and safety rating system operating guidance: Addendum for the profile for the hazard of fire and in relation to cladding systems on high rise residential buildings.* Online. Available: https://assets.publishing.service.gov.uk/government/uploads/system/uploads/attachment_data/file/760150/Housing_Health_and_Safety_Rating_System_WEB.pdf (accessed 31 October 2021).

Ministry of Housing, Communities and Local Government. (2019). *Outcomes of report on Housing Health and Safety Rating System (HHSRS) scoping review.* Online. Available: www.gov.uk/government/publications/housing-health-and-safety-rating-system-outcomes-of-the-scoping-review/outcomes-of-report-on-housing-health-and-safety-rating-system-hhsrs-scoping-review (accessed 15 September 2021).

Ministry of Housing, Communities and Local Government. (2020a). *English housing survey energy report 2019–20.* Online. Available: https://assets.publishing.service.gov.uk/government/uploads/system/uploads/attachment_data/file/1000108/EHS_19-20_Energy_report.pdf (accessed 31 October 2021).

Ministry of Housing, Communities and Local Government. (2020b). *English housing survey headline report 2019–20.* Online. Available: https://assets.publishing.service.gov.uk/government/uploads/system/uploads/attachment_data/file/945013/2019-20_EHS_Headline_Report.pdf (accessed 5 October 2021).

Nearly Legal. (2012). *The cold renormalisation*. Nearly Legal Blog Housing Law News and Comment. Online. Available: https://nearlylegal.co.uk/2012/06/the-cold-renormalisation/ (accessed 25 November 2021).

Office of the Deputy Prime Minister. (2006). *Housing Health and Safety Rating System Operating Guidance*. Online. Available: https://assets.publishing.service.gov.uk/government/uploads/system/uploads/attachment_data/file/15810/142631.pdf (accessed 25 September 2021).

Office of the Deputy Prime Minister. (2006). *Reducing the risks: The housing health and safety rating system*. Online. Available: https://assets.publishing.service.gov.uk/government/uploads/system/uploads/attachment_data/file/8262/144527.pdf (accessed 3 October 2021).

Ormandy, D. and Battersby, S. (2020). The HHSRS scoping review: What a dog's breakfast. *Journal of Housing Law*, 23(1), pp. 107–110.

Scottish Government. (2018). *Tolerable standard*. Online. Available: www.mygov.scot/landlord-repairs/tolerable-standard (accessed 31 October 2021).

Shelter Scotland. (2021). *The tolerable standard*. Online. Available: https://scotland.shelter.org.uk/professional_resources/legal/housing_conditions/the_tolerable_standard (accessed 31 October 2021).

Universities of Kent and Bristol. (2017). *Closing the gaps: Health and safety at home*. Online. Available: https://assets.ctfassets.net/6sxvmndnpn0s/7H2fzdBc0LyHBZDzYuRwA8/54ac1313c52e3e912fced09e69463844/2017_11_14_Closing_the_Gaps_-_Health_and_Safety_at_Home.pdf (accessed 31 October 2021).

9 Regulating houses in multiple occupation (HMOs)

Louise Harford and Kevin Thompson

Introduction

HMOs are essentially low-cost, non–self-contained accommodation. They include one-room lettings with shared amenities such as kitchens, bathrooms or WCs, non–self-contained accommodation, hostels and similar. So-called Section 257 (s257) HMOs are a specific type of HMO which contain poorly converted multiple self-contained units. The different HMO types are discussed in detail in what follows. HMOs meet specific housing needs which may vary depending on where they are located. They are regulated differently from other types of housing, as they present particular housing risks and often contain poor housing conditions or management. HMOs also present particular risks around fire safety, overcrowding and over-occupation and can be immensely challenging to regulate.

Legislative context

The legislative landscape for HMOs is predominantly contained in Parts 1 to 3 of the Housing Act 2004. Part 1 sets out a system for assessing and enforcing housing conditions. This applies to HMOs as for other housing and is considered in Chapter 8 of this book. Part 2 concerns HMO licensing schemes and Part 3 selective licensing schemes, which can apply to HMOs as discussed below. There are also planning tools available which focus on the class of use.

Definition of houses in multiple occupation

The Housing Act s254 defines a building or part of a building (e.g. a flat) as a house in multiple occupation (HMO) if it meets one of three 'tests':
   The standard test – a building or part of a building which:

(a)   consists of one or more units of living accommodation not consisting of a self-contained flat or flats;

DOI: 10.1201/9781003246534-10

(b)   the living accommodation is occupied by persons who do not form a single household (defined under s258);
(c)   the living accommodation is occupied by those persons as their only or main residence or they are to be treated as so occupying it (defined under s259);
(d)   their occupation of the living accommodation constitutes the only use of that accommodation;
(e)   rents are payable or other consideration is provided in respect of at least one of those persons' occupation of the living accommodation; and
(f)   two or more of the households occupying the living accommodation share, or the accommodation is lacking in, one or more basic amenity (toilet, personal washing facilities or cooking facilities).

The self-contained flat test – defined as a self-contained flat which meets the above criteria (b) to (f);

The converted building test – defined as a converted building with one or more units of living accommodation that do not comprise a self-contained flat and meets criteria (b) to (e).

In addition, an HMO may be defined by virtue of being declared an HMO by the LA (under s255). This power is mostly used where there is uncertainty as to whether the property falls within any of the s254 or 257 tests. An example might be a part of a building used as permanent staff or worker accommodation. HMO declarations can be challenged at the First-Tier Tribunal (Property Chamber).

Section 257 HMOs are often referred to as "poorly converted flats", are a specific type of HMO not defined under any of the s254 tests. These are converted blocks of flats in which the standard of conversion does not meet, at a minimum, that required by the 1991 Building Regulations and less than two-thirds of the flats are owner occupied.

The meaning of 'household' in the tests is defined in s258 of the act as follows:

•   Families (including single persons) and co-habiting couples (whether or not of the opposite sex), or,
•   Other relationships as defined in SI 2006:373 regulation 3, e.g. employees, domestic staff, fostering or carer arrangements.

Any other living arrangement e.g. house or flat shares, housing clubs etc. are not households for the purpose of the definition and will not exempt the property from HMO status.

HMOs: fire safety, crowding, over-occupation and management

Fire safety is concerned with means of escape in case of fire and other fire precautions and should be considered with regard to Local Authorities Coordinators of Regulatory Services (LACORS) guidance (LACORS, 2008). This is currently subject to review in part relating to the Grenfell tragedy and subsequent findings in the Hackitt Report in 2018 (Hackitt, 2018) and the wider HHSRS review (see Chapter 8).

Crowding and space (or over-crowding) – put simply – is concerned with space and rooms available for the number of occupants. There are a range of provisions around this, and this is further outlined in Stewart and Lynch (2018). Conversely over-occupation is concerned with the number of amenities – WCs, wash hand basins and kitchen amenities available per person occupying the HMO and their situation. These can be dealt with under Part 1 of the Housing Act 2004 (HHSRS) but are more often dealt with by licencing and, in relation to fire safety, reference is made to LACORS guidance (LACORS, 2008).

The Management of Houses in Multiple Occupation (England) Regulations 2006 provide specific remedy for management issues, placing duties on both managers and tenants to ensure proper maintenance and to generally act responsibly. There are requirements to provide the manager's availability details in case of emergency and to help protect occupiers from injury, safety of gas and electric supplies and a duty to maintain the interior as well as to ensure proper refuse storage. More detail is provided in Stewart and Lynch (2018).

HMO regulation and licensing

HMOs can be regulated through property licensing schemes which aim to manage the quality, safety and management of HMOs (also refer to Chapter 10). Licensing requires that the landlord meets prescribed requirements in order to obtain a licence. HMO licensing is not concerned with the location or density of HMOs but rather with the safety and management of the property. Under Parts 2 and 3 of the act there are three types of licensing which may be utilised by an LA to regulate HMOs.

Mandatory licensing (Part 2): This scheme applies nationally and covers the following types of HMO containing five or more persons across two or more households, regardless of storey count:

- HMOs meeting the standard test
- HMOs that are flats in non–purpose-built blocks
- HMOs that are flats in purpose-built blocks containing no more than two such flats.

In the case of flats this includes those above or below commercial premises.

Mandatory licensing does not apply to s257 HMOs.

Additional licensing (Part 2): Local authorities may designate areas of additional licensing, which extends the mandatory licensing scheme to apply to a wider range or all HMOs in part or all of its district. They must demonstrate a need to improve the quality and safety of HMOs or their surrounding environment e.g. due to poor management. To do this, they must follow the procedures set out in Part 2 of the Act, sections 56–60. Additional licensing may apply to s257 HMOs.

Licences under both schemes must contain conditions as to minimum property standards and management. Schedule 4 to the act as amended (SI 2018:616) sets out mandatory conditions that must be applied, but local authorities may apply other conditions they consider appropriate in relation to the management, use, occupation, condition and contents of the house.

Selective licensing (Part 3): Local authorities can designate part or all of their area as subject to selective licensing. All private rented housing in the designated area apart from those already covered by either of the two HMO schemes outlined earlier would be covered by the scheme. Selective licensing will therefore only be relevant to HMOs where a comprehensive additional licensing scheme is not in place. An additional licensing scheme will be more effective in regulating HMOs than a selective licensing scheme unless the prime focus is on anti-social behaviour or crime. For a selective licensing designation, the area must meet criteria relating to at least one of the following:

- Poor housing conditions
- Low housing demand, or is likely to become such an area
- A significant and persistent problem caused by anti-social behaviour (ASB)
- High levels of migration
- High level of deprivation
- High levels of crime.

In designating a selective licensing scheme, the LA must follow the procedures set out in Part 3 of the Act, sections 80–84.

Failure to comply with the provisions of a licensing scheme is a criminal offence under s72 or s95 punishable on summary conviction by an unlimited fine and failure to comply with a licence condition, a level 5 fine (or Civil Penalty Notice up to £30,000).

Exemptions from licensing

Certain categories of property are exempt from the HMO definition and therefore HMO licensing. These are set out in Schedule 14 to the Act and SI 2006: 373. These include buildings occupied by only two persons in two households, buildings owned and managed by certain public bodies or regulated other than under the Housing Act 2004. A property with a resident landlord with no more than two lodgers in addition to their family is also exempt (SI 2006: 373, regulation 6). Powers under s79(4) of the act exempt certain premises and tenancies from selective licensing. These are set out in SI 2006: 370.

Relationship between Part 1 enforcement and licence conditions in respect of hazards

Under s 67(4) of the act local authorities must seek to identify, remove, or reduce category 1 or category 2 hazards in the house by the exercise of Part 1 functions and not by means of licence conditions. This does not, however, prevent the authority from imposing licence conditions relating to the installation or maintenance of facilities or equipment even if the same result could be achieved by the exercise of Part 1 functions. The imposition of a licence condition for any purpose does not affect the operation of Part 1 functions.

A tale of two postcodes: Liverpool and Hackney

Liverpool

Liverpool is a city and metropolitan borough in Merseyside, England. Over recent years, population growth in the city has increased steadily, with the Office of National Statistics (ONS) population estimate for Liverpool in 2018 being 494,800. This makes Liverpool one of the most heavily populated cities in the UK, with further population growth expected over the next two decades. Recently, Liverpool has shown significant inward migration in the 15- to 24-year age group, driven in part by increased student populations (ONS, 2018). Liverpool has a high proportion of high-deprivation wards, and the city is ranked number two for deprivation nationally, with Liverpool having a significantly higher proportion of residents in fuel poverty (17%) than the national average (11.1%).

The private rented sector now represents 32% of all households (circa 70,000 homes) with social rented households at 44% and 24% in owner-occupation.

A council's property age profile can have an impact on housing conditions, and Liverpool has a high proportion of its residential housing stock built pre-1900 (26.7%) with the majority of Liverpool's stock built before the Second World War (54.2%). There are 17,500 private rental properties in Liverpool that are likely to have a serious home hazard. Liverpool has 16,494 HMOs, 2,882 of which are licensable under part 2 of the act. Regulating the PRS reactively is resource intensive, and Liverpool made 12,172 interventions in

PRS properties over a 5-year period, resulting in 7,844 housing and public health statutory notices and 1,907 prosecutions (Metastreet, 2020).

Planning and HMOs

The planning definition (seven or more occupants) of an HMO differs from that of the Housing Act definition, and this is important when considering the term of HMO licences. One of the key pieces of case law in recent years (*London Borough of Waltham Forrest (LBWF) v Khan*, 2017) discusses the issuing of shorter licences for selective licensing under Part 3, and the outcome also affects HMOs within the designation and Mandatory licensable HMOs under Part 2. After an initial decision, LBWF subsequently appealed and won at the Upper Tribunal (Lands Chamber), where Martin Rodger QC, the Deputy Chamber President, stated: "it was unnecessary and unrealistic to regard planning control and Part 3 licensing as unconnected policy spheres in which local authorities should exercise their powers in blinkers". It is therefore essential that local authorities do have regard to planning status when proposing to grant licences. In addition, a written policy on factors taken into consideration when determining the length of a licence is vital in the spirit of transparency and fairness.

Cumulative impact of HMOs

Whilst HMOs do provide an affordable tenure for citizens, large-scale proliferation of HMOs can damage neighbourhoods. In light of permitted development rights a single dwelling can be converted to a 'small HMO' of up to six persons without planning permission and vice versa. The implication of this permitted development right for local authorities is that the development of smaller HMOs is generally not visible to the planning department and cannot be regulated in planning terms.

Liverpool has been impacted by the proliferation of HMOs in specific neighbourhoods in recent years and has had to take a strategic approach to managing the issue by introducing proactive planning controls under Article 4 Direction due to a steady rise in smaller HMOs uncontrolled by planning legislation. This means all HMOs within a designated area must apply for planning permission regardless of their size. The cumulative impact of HMOs is disrupting neighbourhoods. For example, in one street in Liverpool there are 31 licensed HMO properties that make up 58% of the overall PRS on the street. The impact on the transient nature of PRS properties, particularly HMOs, is evident within this neighbourhood in relation to waste management issues, anti-social behaviour in addition to traffic management and parking concerns.

In addition, average house prices in Liverpool (£125,000) are comparatively low compared to the national average (£240,000), making Liverpool an attractive city for buy-to-let landlords to convert properties into HMOs who are perhaps not as invested in neighbourhoods as homeowners.

HMO regulation in practice: a case study from Liverpool

How we became involved

The Healthy Homes Team raised concerns about infestations and overcrowding.

The property

The building comprises a conversion into four self-contained flats situated over a commercial store. The age of the conversion is unclear; internal checks with planning and building control identified no building regulation application to the City Council or planning permission. A formal request for the completion certificate via a Housing Act 2004 s235 notice was served and not returned. The City Council believes the building satisfies the test under s257 of the act.

Who is the landlord?

The council encountered problems identifying the landlord, and the tenant was unable to assist with information requests due to language barriers.

A Requisition for Information under The Local Government (Miscellaneous Provisions) Act 1976 s16 was served on the freeholder, who subsequently provided the details. Appropriate notice was given under Housing Act 2004 s239 to enter.

Inspections

Flats 1, 3 and 4 were inspected along with the common parts of the HMO, and multiple hazards were found in all flats and over-crowding in flat 4, see Figure 9.1. Within each flat 14 hazards were identified, including Damp and mould growth, Excess cold, Crowding and space, Entry by intruders, Domestic hygiene, Pests and Refuse, Food safety, Personal hygiene, Sanitation and Drainage, Falling between levels, Falls on the stairs, Electrical hazards, Fire, Domestic hygiene and others.

Outcome (ongoing)

This is ongoing at the time of writing but includes: £700 in charges under s49 of the act, referral to food safety team, warrant to enter flat 2, property improved with social value added as local contractors used. Enforcement of

Figure 9.1 Photograph of kitchen in HMO

common parts issues pertaining to s257 HMOs and under The Licensing and Management of Houses in Multiple Occupation and Other Houses (Additional Provisions) (England) Regulations 2007.

HMO regulation in practice: a case study from Hackney

Hackney is an inner London Borough situated just north of the City of London with a population of 281,100 (ONS, 2018). It is currently the third-most-densely-populated LA in the UK. Hackney has a relatively young population with around 25% being under 20 and only 15% over 55. It is culturally diverse, with just over a third (36%) of respondents to the 2011 Census in Hackney describing themselves as White British (see also London Borough of Hackney Policy and Insights Team, 2020)

The private rented sector now represents 30% of all households (33,923 homes), with social rented households at 44% and 26% in owner occupation. Hackney is ranked the 22nd-most-deprived LA in the England (2019 Index of Multiple Deprivation. 15% of private rented homes are overcrowded, and 11% contain Category 1 hazards, this rises to 20% for HMOs. There are an estimated 4315 HMOs in the borough, and of these, 77% fall outside the Mandatory Licensing Scheme criteria (BRE, 2017).

Affordability of housing in Hackney is a major challenge. The average purchase price for a one-bed flat is £428,000 (HM Land Registry June, 2021). The average rent for a one-bedroom flat in the PRS is £1,460, £1,725 for a two-bedroom (ONS, 2018). HMOs therefore provide a valuable supply of affordable housing, and for many people it is the only option. House and flat sharing is common among the younger age group. Considering the challenges, on 1 October 2018 Hackney introduced a Borough-wide additional licensing scheme covering all HMOs not covered by the Mandatory scheme (estimated 3324 HMOs). It also introduced a pilot selective licensing scheme for all other PRS homes in three wards where data showed the highest prevalence of Category 1 hazards and disrepair are to be found (BRE, 2017).

How we became involved

Referral from neighbourhood police officer following a visit to the premises with the London Fire Brigade which identified two units of living accommodation believed to be multiply occupied. See Figure 9.2.

The property

There were allegations that people were living in an outbuilding to the rear yard of a car wash and tyre depot. The main part of the premises was located in a railway arch. The building had planning permission only for commercial use. No HMO licence application had been made.

An initial 'walk-by' visit could not confirm whether the outbuildings were occupied, but it was clear the outbuildings were not designed to be used for housing. We found one single-storey structure adjacent to a two-storey structure; the construction was seen to be poor:

- walls appeared to be a single skin of brick/breeze block/ply boarding
- flat and shallow pitched roof covered with roofing felt
- small and insufficient windows indicating inadequate natural lighting and ventilation.

Figure 9.2 Exterior of HMO

An early-morning 'raid' under warrant of entry was carried out with police and a locksmith. This confirmed the two-storey structure was an HMO containing three lettings. The single-storey structure was similar to a studio flat but lacking windows and fixed heating and completely unsuitable for housing.

Findings

The HMO was used as storage for the business with tyres piled high in the common parts and in one of the bedrooms. Conditions were as follows:

- Occupied by 4 people in 3 households
- Large accumulation of flammable tyres partially blocking the communal stairway and landing, impeding the escape route and would have emitted toxic smoke in the event of a fire
- Electrics were poor and dilapidated
- Stairs to first floor were steep and in disrepair
- Only one battery smoke detector that had been disabled
- No door to communal kitchen, so the means of escape would be flooded with smoke if a fire occurred there; the partition separating the kitchen from the bedroom was of ply boarding
- No hot water to sink or washbasin
- Concerns about domestic water supply provided from a tank in a very dirty warehouse
- Two of the three bedrooms lacked windows.

Outcome

There was a successful conviction at Thames Magistrates Court. The defendant pleaded guilty to operating an HMO property without a licence and to five breaches of The Management of House in Multiple Occupation (England) Regulations 2006, resulting in a £8,992 fine. Residential use of the site has ceased.

References and legislation

Building Research Establishment (BRE). (2017). *BRE integrated dwelling level housing stock modelling and database for hackney council; Report Number: P104088–1009 18 April 2017.* Watford: BRE.

Hackitt, J. (2018). *Independent review of building regulations and fire safety: Final report.* London: MHCLG.

HM Land Registry (2021). Land Registry Search https://www.land-search-online.co.uk/land-registry/?i=land-registry-search&gclid=Cj0KCQiA8vSOBhCkARIsAGdp6RQVU60p2OOO3EaYXMvBZsQYmJFkQEehoiz8egzRpx2hy8z8ftSy7eoaAvxYEALw_wcB

LACORS. (2008). *Housing fire safety: Guidance on fire safety provisions for certain types of existing housing.* London: LACORS.

The Licensing and Management of Houses in Multiple Occupation and Other Houses (Miscellaneous Provisions) (England) Regulations 2006 SI 2006: 372.

The Licensing and Management of Houses in Multiple Occupation and Other Houses (Additional Provisions) (England) Regulations 2007 SI 2007: 1903.

The Licensing of Houses in Multiple Occupation (Prescribed Description) (England) Order 2018 SI 2018: 221.

London Borough of Hackney Policy and Insight Team. (2020). *A profile of hackney, its people and place, London borough of hackney policy and insight team*, Online. Available: https://drive.google.com/file/d/1JZLZFzNUSO4017-vCA_dy9Dk08e6jXa_/view?usp=sharing (accessed 25 November 2021).

London Borough of Waltham Forrest (LBWF) v Khan, 2017 UKUT 153 (LC) https://www.bailii.org/uk/cases/UKUT/LC/2017/153.html

The Management of House in Multiple Occupation (England) Regulations 2006 SI 2006: 372.

Metastreet. (2020). Liverpool city council private rented sector: Housing stock condition and stressors report July 2020. *Metastreet.*

ONS. (2018). *Mid-year population estimates for major towns and cities, 2016.* Online. Available: www.ons.gov.uk/peoplepopulationandcommunity/populationandmigration/populationestimates/adhocs/008264midyearpopulationestimatesformajortownsandcities2016 (accessed 20 October 2021).

The Selective Licensing of Houses (Specified Exemptions) (England) Order 2006 SI 2006: 370.

Stewart, J. and Lynch, Z. (2018). *Environmental health and housing* (2nd ed.). Oxon: Routledge.

10 Housing Act 2004 property licensing schemes

Henry Dawson and Richard Tacagni

Introduction

Housing Act 2004 property licensing schemes build upon the powers the act bestows upon LAs to help them deal with general safety and maintenance issues in private rented properties. Licensing is mainly used for large HMOs with five or more occupiers (in Wales these must also be three or more storeys), but it can also be brought in over a designated LA area to address problems with smaller HMOs or non-HMO properties. These discretionary, area-based schemes have become very popular with LAs (Chartered Institute of Housing and Chartered Institute of Environmental Health, 2019; National Landlords Association, 2015; Wilson, 2015).

The Housing Act 2004 allows LAs to license individual privately rented properties, imposing conditions on their licences regarding levels of occupation, amenities (e.g. bathrooms, fire safety, ventilation, etc.), management processes (e.g. providing written tenancy agreements, handling anti-social behaviour, etc.), and a requirement for people to be fit and proper persons to be able to manage those properties. Licensing provides an additional layer of controls for private rented properties.

Licensing seeks to impose threshold standards for properties housing the poorest and most vulnerable in society (Rooker, 2004). Weak security of tenure and high demand for accommodation lead to under-reporting of problems with rental properties, particularly in the lower-cost areas of the market (de Santos, 2012; Lister, 2004). Licensing is a proactive measure dealing directly with landlords and agents, so it does not rely on tenants making the council aware of problems in their homes before enforcement action can commence. It is self-financing through charging licensing fees, providing a system in which the market largely pays for this area of its governance.

Giving LAs the discretion to set their own licensing conditions also allows schemes to be tailored to reflect regional variation in the PRS markets (Moffatt and Watson, 2018). With the introduction of discretionary licensing LAs

DOI: 10.1201/9781003246534-11

can introduce additional and selective licensing in part or all of their area to bring about market change and (in collaboration with a range of partners) introduce PRS-led regeneration and renewal activity. Finally, allowing LAs to administer licensing schemes allows them to improve data on their local markets (Lawrence and Wilson, 2019).

Setting up a new scheme

Developing an additional or selective licensing scheme is a major undertaking that requires time, resources, and political support. It is not something that can be rushed. A time scale of around 12 to 18 months should be expected.

Delivering a new licensing scheme requires a multi-disciplinary project team that reports to the council's senior leadership team, cabinet, and council leaders. Setting clear aims and objectives will help to steer the project. For example, what are the issues the council are seeking to address, what is the evidence base, and how will licensing (when combined with other interventions) help to address the issue?

Early engagement with internal and external stakeholders can be useful during the formulative stage, sharing thoughts and ideas and gaining feedback as the scheme proposal evolves. It is never too early to start the engagement process.

Accompanying scheme development should be a review of the council's HMO standards, licence application process, and proposed licence conditions. In 2018, the Court of Appeal ruled that selective licence conditions are restricted to 'management, use and occupation', whereas mandatory HMO and additional licence conditions can also extend to 'condition and contents' (*Brown v Hyndburn Borough Council*, [2018] EWCA Civ 242). Conditions need to be drafted accordingly. Plus all licences must contain the prescribed conditions in Schedule 4 of the Housing Act 2004 (as amended). LAs imposing conditions on their licences outside of these criteria are acting *ultra vires* and run the risk of legal action and the potential for expensive remedial works and civil court damages as a result of forcing property owners to carry out works to their property which the LA was not legally entitled to require them to complete.

Locally determined licence conditions allow schemes to be tailored to local property markets, area-based issues, and construction features in the local housing stock, but they lead to wide variation in the conditions being imposed by neighbouring authorities. This can lead to confusion when landlords have portfolios across a number of LA areas. The only nationally available guidance on setting licence conditions was provided in the form of a set of suggested licence conditions set out by Shelter (Murphy and Mitchell,

2006). There is a need for central government direction and support when setting licence conditions (Chartered Institute of Housing and Chartered Institute of Environmental Health, 2019). Until this is available neighbouring authorities should collaborate to develop agreed sets of conditions with minimal local variation and seek to emulate best practice, looking to similar authorities with established and successful schemes.

Before approving the scheme, the council must take 'reasonable steps' to consult everyone likely to be affected by the designation (Housing Act 2004 s56(3) & 80(9)) and must consult for at least 10 weeks (Ahmad, 2015). In 2015, the High Court held that 'reasonable steps' rather than 'all reasonable steps' implies comparatively wide discretion as to how the consultation process is carried out when dismissing a judicial review against Croydon Council's selective licensing scheme (*Croydon Property Forum Ltd, R [on the application of] v The London Borough of Croydon* [2015] EWHC 2403).

The consultation phase is critical to successful delivery of a new licensing scheme, and nothing should be left to chance. It requires a comprehensive consultation report and supporting evidence base, effective consultation arrangements, and a comprehensive communications plan.

Deficiencies in the consultation process can result in legal challenge through the judicial review process. In 2014, Enfield Council's additional and selective licensing schemes were quashed by the High Court (*R [Regas] v LB Enfield* [2014] EWHC 4173) after it was held the council had not consulted widely enough or for the required period. Other councils have stepped back from scheme implementation or re-run their consultation process when deficiencies have been highlighted. Effective planning can help to design out or at least minimise such risks.

Following completion of the consultation, the council must consider all representations made before deciding whether to approve a licensing scheme. The decision is usually made at a cabinet or full council meeting, although arrangements can vary according to the council's constitution.

Since 2015, selective licensing schemes including more than 20% of private rented homes in the borough and/or 20% of the geographical area of the borough require Secretary of State approval (Ahmad, 2015). This can involve a lengthy application process, and the outcome is not guaranteed. For example, Liverpool, Hastings, and Croydon Councils all had selective licensing schemes rejected by the Secretary of State, whereas Barking and Dagenham, Waltham Forest, and Enfield Councils all had selective licensing schemes approved by the Secretary of State.

Implementing an additional or selective licensing scheme involves making a scheme designation at least 3 months before the scheme comes into force. It is advisable to seek legal advice when drafting the designation, as errors can be difficult to resolve. Once a designation has been made, councils can revoke a designation following a review. To date, the courts have not ruled on whether a designation can be varied or 'partially revoked'. Pending an affirmative court decision, this approach is not recommended.

Before making the scheme designation, it is vital to ensure all the building blocks of the scheme are in place. The application system should be up and running so that landlords can submit licence applications before the scheme comes into force. For larger-portfolio landlords and agents, this can be a time-consuming task undertaken during the three-month lead-in period.

Whilst ignorance of the law is no defence, it is incumbent on councils to widely promote any licensing scheme and bring it to the attention of land-lords and agents, both within and outside the local area. Publicity through the council's website and social media channels, landlord forums, landlord newsletters, articles, and adverts in the trade press and promotion via profes-sional lettings industry bodies can all help to spread the word.

Licence applications and fees

Councils can charge a fee for licence applications (Housing Act 2004 s63(3) & 87(3)), but they are not able to charge for temporary exemption notices or variation of existing licences. Administering licensing schemes is an expen-sive and long-term commitment. LAs must predict administrative costs from licence application processing and property inspections alongside projected costs from addressing non-compliance and enforcement work. The final fee structure must balance what their projected costs are for the typical five-year period of a licence against what is considered to be 'locally acceptable' as a total cost for a property licence. If licence fees are too high landlords may choose to avoid applying, greatly increasing the costs involved in locating and pursuing unlicenced operators (often in numbers well beyond the capac-ity of LA enforcement and legal departments). If fees are too low, then there will be inadequate resources to proactively pursue non-compliant landlords. At best this can lead to resentment from those who have complied with the scheme; at worst this can reduce the licensing scheme to what amounts to 'an expensive paper exercise' (Lawrence and Wilson, 2019: 56).

In 2017, the Divisional Court ruled that licence application fees must be paid in two parts (*Gaskin, R [on the application of] v Richmond Upon Thames London Borough Council & Anor* [2017] EWHC 3234). The Part 1 fee is restricted to costs incurred in processing the licence application. If the licence is to be granted, the Part 2 fee is payable, covering the cost of administering and enforcing the licensing scheme.

Careful planning is needed when setting licence application fees if the intention is to make the scheme self-funding. In 2021, all selective licensing fees in London were a fixed fee per property. Whilst other fee models have been considered, such as a variable fee based on council tax banding, they are often dropped as being too complicated.

HMO licensing fees vary from a fixed fee per property, a baseline fee combined with an extra fee for each bedroom, or a fixed per-bedroom fee with no baseline fee. There are pros and cons of each approach. Generally, the more complicated the fee structure, the more difficulty it creates when processing applications.

Integrated online application and payment systems can help to reduce costs associated with administering a licensing scheme. This can free up resources to undertake a risk-based inspection programme for licensed properties, combined with an enforcement programme targeting landlords who seek to evade the scheme.

Fee discounts are worth considering. For example, an 'early-bird' discount can help maximise applications submitted when a new scheme is implemented. Accreditation discounts for the licence holder and designated property manager can help encourage higher standards of property management. Energy efficiency discounts can incentivise investment to achieve Energy Performance Certificate (EPC) Band C or above.

Licensing schemes cannot make a profit or be used to cross-subsidise other service areas. To avoid this happening, councils must retain flexibility to adjust staffing levels to keep the scheme on budget and maintain tight financial control.

There has been a considerable uplift in licence application fees in recent years. For example, when Newham introduced borough-wide additional and selective licensing in 2013, they charged a standard licence application fee of £500 per property, reduced to £150 with an early-bird discount. This equated to just 60p a week and was tax deductible (Moffatt, 2016).

In London (April 2021), the average mandatory HMO licensing fee for a five-bedroom HMO was £1,292, the average additional licensing fee for a four-bedroom HMO was £1,161, and the average selective licensing fee was £640. In London, average mandatory HMO licensing fees have increased by 43% over the last six years (Tacagni, 2021). If fees continue to rise, combined with the roll-out of HMO Article 4 Directions, this has the potential to adversely affect the supply of lower-cost shared accommodation.

Inspections and enforcement

Granting a licence based on a desktop assessment will not in itself improve housing conditions or address poor management practices. With mandatory HMO and additional licensing, councils have a duty to satisfy themselves that there are no Part 1 functions that ought to be exercised (regarding housing-related hazards) as soon as possible, and certainly within five years of the licence application date (Housing Act 2004 s55(5)(c) & (6)(b)).

With selective licensing, if the scheme was implemented to address poor housing conditions, there is a requirement to inspect 'a significant number of the properties' within the designated area (The Selective Licensing of Houses (Additional Conditions) (England) Order 2015 SI 2015/977).

To satisfy these obligations and help achieve scheme objectives, a proactive inspection programme should be developed. Each council will do this differently, and some may not inspect all licensed properties. For example, rather than ask for routine submission of documents, Newham Council listed all the requirements in licence conditions. They undertook audit spot checks to ensure compliance. Licence holders unable to provide a full suite of safety documentation were in breach of the licence, and inspections could be arranged to pursue the investigation.

Operating at arms' length from licence application processing, suitably experienced officers can be tasked with enforcing the scheme in accordance with local policies and procedures. Experienced enforcement officers can focus on the most complex investigations, working in a multi-agency capacity to address the most serious cases. Armed with complementary enforcement powers, Trading Standards Officers can help to tackle offences involving letting and managing agents alongside the licensing scheme (National Approved Letting Scheme, 2018). When enforcement is combined with the requirement for licence holders and property managers to be fit and proper persons, licensing schemes provide the tools to drive poor operators out of the market completely (Lawrence and Wilson, 2019).

Searches for unlicensed properties have become more sophisticated. Detailed desktop work using a range of national and local data is used to inform more labour-intensive measures such as door-knocking and inspections using powers of entry. Close liaison and information sharing with housing benefit and council tax departments provides reciprocal benefits through identification of unlicensed properties and uncovering of evidence of fraud and non-payment of taxes.

Focused and persistent action on non-compliant properties is important for the success of licensing schemes. In one London borough this helped to increase selective licence applications from 27,000 to 49,000 over a five-year period (Lawrence and Wilson, 2019). Whilst hundreds or even thousands of properties will potentially remain unlicensed, wide promotion of enforcement work presents a legitimate threat to non-compliant operators and can be partially funded through charging increased fees for late licence applications. Investigating and prosecuting hundreds of landlords is possible (see Newham Council's licensing work) but requires levels of resources which are unavailable to most licensing departments. All authorities should be undertaking at least some well-publicised legal activity against non-compliant operators. Variation in levels of enforcement is one of the main criticisms of licensing schemes (National Landlords Association, 2015; National Residential Landlords Association, 2021).

Evaluating licensing schemes

Councils must review their additional and selective licensing designations from 'time to time' (Housing Act 2004 s60(3) & 84(3)). The frequency is undefined. Establishing baseline data, SMART targets, and objectives can help with keeping the project on track, using a variety of quantitative and qualitative measures. Not only will this help to demonstrate the scheme is being effectively implemented, it also highlights issues that need to be addressed during the operation of a licensing scheme.

Regular updates can be shared through landlord forums, newsletters, press releases, and other such mechanisms. These can explain licensing scheme performance, promote licensing activity, and emphasise the positive effects it is having using data and case studies. Licensing schemes can be subject to critical news and social media reporting. They can appear to landlords as a form of extra taxation and to tenants as a cause of rent increases. LAs must publicise the effectiveness of their schemes to win political and other support for the continuation and acceptance of licensing as a mechanism of improving housing conditions across an area.

Clear mechanisms for evaluation of schemes should be put in place at the design stage, not applied *post hoc*. When designing evaluation, it is important to differentiate between measurements of outputs which can easily be quantified (e.g. the number of properties inspected for housing-related hazards) and the contribution towards the overall outcomes for the scheme (e.g. improvement in the housing stock or improvement in the health of residents). Many of the benefits of licensing, especially those regarding area-based improvements, may take longer than the five-year period of a typical property licence to be fully realised (Lawrence and Wilson, 2019). There are long lead-in times as inter-agency working and effective collaborative relationships are formed, but this aspect of licensing work leads to many of the wider benefits of licensing schemes and must be considered in the evaluation of the effectiveness of those schemes.

Lawrence and Wilson's (2019) study of selective licensing attempted to measure the impact of licensing on indices of multiple deprivation across selective licensing areas. Their findings were inconclusive, highlighting how the lack of comparison data over the whole or part of individual Lower Super Output Areas makes it difficult to measure outcomes of schemes directly.

LA belief in the effectiveness of licensing has increased significantly since the Building Research Establishment's national study of licensing in 2010. In 2019 Lawrence and Wilson found that 92% of LA survey respondents saw licensing schemes as an effective means for dealing with property conditions and anti-social behaviour. They also found that effectiveness of schemes was greatly increased when they were integrated into wider LA strategies and coherent, well-planned and -resourced initiatives to improve conditions in the PRS with appropriate political support.

Home visits by inspectors need not be restricted to the enforcement of licensing conditions. The self-resourced inspection team formed to administer a licensing scheme can be a source of information for a range of services. Referrals can be made to various public health–related bodies, energy efficiency schemes, social services, welfare and benefit support, dedicated charities (e.g. Age Concern), police home security services, fire and rescue services, departments for the detection of modern slavery, sex trafficking, and border control agencies (Lawrence and Wilson, 2019). Monitoring and recording these 'spin-off' benefits of schemes is an important part of demonstrating the wider benefits of licensing.

Conclusions

Licensing schemes provide LAs with a powerful and mostly self-funded mechanism for ensuring threshold standards are implemented in the PRS.

Setting up those schemes is a resource-intensive activity and requires care and balanced decision-making to pitch fees and levels of enforcement activity appropriately. Failure to do this effectively can result in schemes not being approved or being halted by a judicial review. Lack of enforcement can also reduce schemes to what may be referred to as a bureaucratic exercise, placing an increased financial burden on compliant landlords whilst non-compliant operators continue to operate without those burdens.

Licensing schemes are contentious, and LAs may encounter opposition from local landlords and tenant groups. Consultation and communication are important to disseminate information on the benefits of schemes. Effective, structured, and ongoing evaluation should be used to address failings as they occur and capture successes to support future licensing activity.

References

Ahmad, T. (2015). *The Housing Act 2004: Licensing of Houses in Multiple Occupation and Selective Licensing of other Residential Accommodation (England) General Approval 2015.* Available: https://assets.publishing.service.gov.uk/government/uploads/system/uploads/attachment_data/file/418588/General_consent_final__2_.pdf (accessed 18 October 2021).

Brown v Hyndburn Borough Council [2018] EWCA Civ 242.

Chartered Institute of Housing, Chartered Institute of Environmental Health. (2019). *A licence to rent.* London: Chartered Institute of Environmental Health.

Croydon Property Forum Ltd, R (on the application of) v The London Borough of Croydon [2015] EWHC 2403.

de Santos, R. (2012). *A better deal towards more stable private renting.* London: Shelter.

Gaskin, R (on the application of) v Richmond Upon Thames London Borough Council & Anor [2017] EWHC 3234.

Housing Act 2004. HMSO: London.

Lawrence, S. and Wilson, P. (2019). *An independent review of the use and effectiveness of selective licensing.* London: Ministry of Housing, Communities and Local Government.

Lister, D. (2004). Young people's strategies for managing tenancy relationships in the private rented sector. *Journal of Youth Studies*, 7, pp. 315–330. https://doi.org/10.1080/1367626042000268944

Moffatt, R. (2016). *Property licensing in Newham [lecture] CEHP.* London: CEHP.

Moffatt, R. and Watson, P. (2018). *Landlord licensing in the private rented sector.* London: Metastreet.

Murphy, J. and Mitchell, S. (2006). *Selective licensing for local authorities.* London: Shelter.

National Approved Letting Scheme. (2018). *Effective enforcement in the private rented sector.* Cheltenham: Safeagent.

National Landlords Association. (2015). *Landlord licensing interim report: Overview of the incidence and cost of HMO & discretionary licensing schemes in England.* London: National Landlords Association.

National Residential Landlords Association. (2021). *The enforcement lottery: Civil penalty usage by local authorities.* Manchester: National Residential Landlords Association.

R (Regas) v LB Enfield [2014] EWHC 4173.

Rooker, J.W. (2004, November 24). HMO licensing: Written statement to the house of lords. *Hansard House of Commons Written Statement*, 657.

The Selective Licensing of Houses (Additional Conditions) (England) Order 2015 SI 2015/977.

Tacagni, R. (2021). *London Property Licensing Fees.* Available: www.londonpropertylicensing.co.uk/ (accessed 18 October 2021).

Wilson, W. (2015). *Selective licensing of privately rented housing (England & Wales) Briefing.* London: House of Commons Library.

11 Advanced regulatory skills and practical evidence gathering

Paul Oatt

Introduction

This chapter examines the importance of gathering and interpreting evidence correctly for enforcement purposes beginning with how to establish who is managing or controlling the property and the reasons we may need to inspect it using our powers of entry. The chapter outlines the type of evidence needed to pinpoint the responsible parties and what to do if a criminal landlord unexpectedly arrives mid-inspection and how evidence is used to support witness statements to overcome any potential lines of defence and make the best possible case. Finally, the chapter deals with the actions available to LAs following successful prosecution for the imposition of banning orders and entry on the Rogue Landlords database.

Gathering evidence before an inspection

Housing enforcement cases often begin reactively in response to a tenants' complaints or arise following proactive routine inspections (Barratt, 2014). Howsoever it may start, evidence must be gathered to justify appropriate enforcement measures or confirm that offences were committed. When serving notices, we must ensure those measures are directed towards the correct persons; otherwise incorrect service of a notice will invalidate it (Stewart and Lynch, 2018).

When evidence is improperly gathered or poorly interpreted, we succumb to 'jumping to conclusions' bias. Investigators must make probabilistic decisions about what the evidence shows (Andreou et al., 2018). To proceed overconfidently with little evidence subconsciously can allow our biases to take over (Huq et al., 1988). It is also important to be wary of the motives of others. Barratt (2014) cautions that motives maybe conscious or unconscious and not to take any claims at face value; the evidence of one's own eyes when inspecting inanimate and motiveless situations, such as the condition of a property, tells the story and will either support or debunk such claims.

DOI: 10.1201/9781003246534-12

Gathering evidence begins before an inspection takes place and starts in the office. LA databases contain numerous evidential sources relating to local properties and residents. The Housing Act 2004 s237 allows officers to use council tax or housing benefit data for investigatory purposes. Information such as land registry searches can be shared between departments (Mishkin and Moffatt, 2013), and LAs publicise planning applications (GOV.UK, 2021). This together with credit searches and historical records of previous enforcement can be used to pinpoint those persons controlling or managing the property. In cases of unlicensed properties, it is customary for LAs to write warning letters to landlords. Enquiries can be used to verify whether or not letters were sent to the owner's current address. Be aware that often non-resident landlords forget to update councils when they move or when appointing agents to manage the property on their behalf. The first step in a licensing case may simply be to get letters sent again to the correct persons or addresses and wait for an application to be made.

Freeholders own the house and the land on which it is built, whilst leaseholders have ownership of property but not land, and then only for a finite term of years after which ownership reverts to the freeholder. Leaseholds therefore are ownerships but also a form of tenancy (Law of Property Act, 1925). Freeholders are generally responsible for external structural maintenance and the common parts and recharge costs to the leaseholders (lessees). Leasehold properties rented to tenants are commonly let by the lessee.

Tenancy agreements are enforceable under contract law whereby an offer is made and accepted between two parties (landlord and tenant) for a consideration going both ways concerning the payment of rent. However, a signed agreement is not always necessary (Contracts (Applicable Law) Act, 1990).

Establishing property control or management depends upon how rent is received. The Housing Act 2004 s263 defines the person having control as the person (owner, freeholder or leaseholder) who either receives rack rent (not less than two-thirds of the premises full net annual value) or would so receive it if the property was let at rack rent. This section also defines a person managing as either an '*owner or lessee*' receiving rents from tenants or licensees either directly or indirectly '*through an agent or trustee*,' who is also considered to be a person managing. What this means is that a non-resident owner of a property whether freeholder, leaseholder, agent or trustee can be a person managing and or in control of a rented property.

Powers of entry and inspection

The Housing Act 2004 s239 covers powers of entry for reactive or proactive purposes. ss239(5) is used to inspect when occupiers complain about poor conditions and landlord negligence towards repairs (reactive). The LA must give at least 24 hours' prior notice. The property is systematically risk assessed internally and externally on its state and condition, noting hazards and deficiencies under the Housing Health and Safety Rating System (HHSRS) (Office of the Deputy Prime Minister, 2006).

The reason for the inspection determines the type of evidence officers need to obtain. In all cases, it is absolutely essential to identify the persons managing and or in control of the property. s239(7) is a proactive approach applicable to the inspection of licensed or licensable properties such as houses in multiple occupation (HMOs) and single-family dwellings designated for selective licensing. Prior notice of entry is not necessary, and officers look to evidence that persons managing and or in control are operating without a licence. s239(7) is also used to routinely inspect licensed properties and determine if the licence holder has upheld or breached licence conditions. HMOs are subject to management regulations. Officers will also inspect under 239(7) to ensure compliance (Housing Act 2004 s72, 95 and 234).

The Housing Act 2004 s239(8) allows enforcement officers to take others with them, enabling a multiagency approach with police, immigration and other partners (Mishkin and Moffatt, 2013). However, despite police presence, entry cannot be forced. Should occupiers refuse access, abandon the inspection and apply to the courts for a warrant under section 240 HA, (2004), making the case to satisfy the magistrates that admission was correctly sought and refused and that application for admission would defeat the purposes of entry. If successful, a warrant allowing forced entry will be granted. Officers should use their pre-inspection evidence in support of the application to prove the property is rented.

When access is gained the full inspection can begin. Any inspection is a snapshot in time. Investigators look for indicators of hazards or offences. Collating and interpreting this information correctly is a core skill (Stewart and Lynch, 2018).

The Housing Act 2004 s239(8) also allows officers to take measurements, photographs or recordings. Recordings include taking a witness statement or officers making notes or sketching out the floorplan. These contemporaneous notes should be made on numbered pages where mistakes are crossed out and not scribbled over. Ultimately there should be no contradictions between what is noted here and what finally goes into an officer's witness statement (Barratt, 2014).

Photographs are taken not only of the living conditions but also occupiers, tenancy agreements and supporting evidence of rental payments through receipts or bank statements (Stewart and Lynch, 2018). Photographs of occupiers' identification may also be used, and production of this identification can be insisted upon when police attend (Mishkin and Moffatt, 2013).

Photographs of bedrooms showing signs of occupancy are important for licensing offences. We need further verification documented through occupiers' witness statements showing how long they have lived at the property and how much rent is paid and to whom. Officers need to confirm if this information matches the intelligence gathered pre-inspection. If it does not, then this produces another line of enquiry.

When interviewing HMO occupiers, it's important to consider the definitions under the Housing Act 2004 s254 and 258 to establish whether occupiers are a family or form at least two households (upwards of three persons for an additional HMO or five for a mandatory HMO) where it is their only or main residence, and they share basic amenities and where at least one occupier pays rent.

Putting it all together

Evidence must be put into exhibits to produce in court, accompanied by witness statements. The process of writing the statement is not dissimilar to preparing a thesis wherein every point made is backed up by a cited reference. In a witness statement one refers to an exhibit as supporting evidence accompanying every stated fact, set out chronologically and avoiding embellishment or opinion (Barratt, 2014). There is no limitation on the number of statements submitted, and they can be written at reasonable intervals where events remain fresh in the memory. For example, a preliminary statement could be made shortly after the inspection covering all the pre-inspection checks you made and the evidence found during inspection showing that offences were committed.

Other lines of enquiry may be discovered, for example occupiers pay rent to an agent. Where there are doubts regarding the management control, further information can be sought through service of a requisition for information under the Local Government Miscellaneous Provisions Act 1976 s16 requesting the recipient to state the nature of their interest in the land, any other parties with interests and occupancy details. Section 235 Housing Act (2004) allows officers to request production of documents reasonably required in the course of housing investigations e.g. tenancy agreements, contracts, inspection records and certifications. Production of this information allows LAs to formally establish details accurately (Stewart and Lynch, 2018). Failure to comply with these notices is punishable by prosecution and fines.

A short witness statement can be made detailing service of the notice and exhibiting it. A follow-up statement can be written up when the information is received back, and these responses are submitted as exhibits. A picture is now being built of who is responsible for the offences committed, the persons managing and or in control of the property.

Memories fade over time, and even though an investigator writes statements using documents, photographs and notes as an aide memoire, the longer time passes between the event and writing the statement, details blur or get forgotten (Barratt, 2014). Human memory and fallibility were tested by Calvillo (2014), showing errors in recollection and cognitive processes arising from misinformation effects wherein events were witnessed and a false fact was subsequently introduced to witnesses during feedback. This falsehood blended with their memories, becoming part of their narratives when recounting events. Similarly, whilst we gather facts we may also be provided with misinformation from suspects, witnesses or complainants whose motivations, conscious or unconscious, serve different agendas.

To reduce potential biases, it is important to detail events in statements, which are signed and dated, showing how closely they were written to the dates of the events depicted whilst being still fresh in your mind.

During inspections of unlicensed properties, landlords may turn up unexpectedly, or it may transpire that a head tenant is subletting the property. By collecting rent and profiteering before paying their own rent, head tenants are also defined as persons managing and or in control. If one is authorised under the Police and Criminal Evidence Act (1984) (PACE) and has sufficiently evidenced the property is operating without a licence and (for HMOs) there are additional management regulation breaches, landlords or head tenants should be interviewed under caution in what is known as a spontaneous PACE interview.

The Police and Criminal Evidence (PACE) 1984 s67(9) provides authorised officers with the same powers as the police to use what suspects say under caution as evidence. It is important to remind suspects that they are not under arrest or obliged to answer (particularly if police are in attendance). Reactions to this maybe hostile, so one needs to keep calm and focus the line of questioning on verifying how they fit the definition of person managing and or in control, so ask them about receiving rent, and if it is known that warning letters were sent, ask landlords to confirm their address. They should be asked them how often they visit the property before asking about why maintenance or repair issues were not dealt with.

This line of questioning is all about countering a reasonable excuse, which is an available defence against housing act offences set out in the Housing Act 2004 s30(4), 32(3), 72(5), 95(4) and 234(4) with no definition of what actually constitutes a reasonable excuse. Housing Act offences become either criminal prosecutions or civil fines. LAs must prove these cases beyond a reasonable doubt (Gardiner, 2017) and prove intent by asking questions that overcome any potential reasonable excuse. Sometimes landlords rely upon arguing that their actions were '*unintentional*' and they honestly believed they were '*doing nothing wrong.*'

In *London Borough of Haringey v Goremsandhu* (2013) it was found that an honestly held belief alone does not constitute a reasonable excuse without there being reasonable grounds for '*holding of that belief.*' All the evidence and witness statements must focus upon proving the offence beyond reasonable doubt and overcoming the reasonable excuse. Ultimately if the excuse is irrefutable, there is no case to answer; conversely a reasonable excuse does not have to be truthful to win a case, the defendant just has to appear credible. But ignoring correctly addressed notices or warning letters, receiving rent and visiting the property are all reasonable indications that a person would be expected to have awareness of their obligations, and it becomes harder to assert an honestly held belief of no wrongdoing.

A good witness statement with strong supporting exhibits helps prove criminal cases without having to attend court, and guilty pleas can be secured during preliminary hearings (Criminal Justice Act 1967 s9). Tenants who provided witness statements may be reluctant or unable to attend court or no longer traceable. Their statements may still be admitted as hearsay under the Criminal Justice Act 2003 s116 where the court is satisfied that witnesses cannot be found, are living abroad or fear repercussions and the defence does not object, although in practice they often do. If tenants' statements are not admitted your statement and exhibits must be strong enough to prove the case. For example, if one's statement says:

> "at the property I met 'John Smith' who allowed access to the ground floor front room telling me and confirming in his witness statement he rents the room for £350pcm sharing it with 'Joe Bloggs' who was not present. I saw two beds, two wardrobes full of clothes, toiletries and other personal possessions, I therefore believe this room is occupied by at least one person, John Smith."

photographs of everything described should be included as well as evidence from John Smith of rental payments from his bank transfer records or tenancy agreement, if he has one. Then if his statement is not included there is sufficient information in one's own evidence, and the case does not solely rely upon the inclusion of statements from others. You will note that whilst as a general rule one should generally refrain from giving opinions it is more acceptable to say "*I believe . . .*" when including supporting evidence.

Banning orders and the Rogue Landlord database

In Chapter 2, Housing and Planning Act 2016, LAs can apply to the First-Tier Tribunal to ban convicted landlords from renting or managing properties in England within 6 months from date of conviction. Once the order is made,

LAs must revoke any property licenses held. Relevant Banning Order offences include Housing Act convictions for non-compliance with improvement and overcrowding notices or prohibition orders, failure to licence HMOs or selective properties, breaches of licence conditions or management regulations and for providing false or misleading information. A ban must last at least 12 months. The tribunal decides the length based on the LA's recommendation.

According to guidance from the Ministry of Housing, Communities and Local Government (2018) the expectation is Banning Orders will be used against the worst offenders dependent upon the severity of the crime. The length of the ban may be influenced by an offender's previous history of non-compliance and their awareness of their own actions and most importantly any harm caused to tenants. Punishment should be proportionate to the offence and long enough to act as a deterrent for future offences and send a message to deter others from doing the same.

The process starts when the LA serves a Notice of Intention to impose a Banning Order. The offender can make representations within 28 days after which time the LA will consider these and decide whether or not to proceed by applying to the tribunal. Breaches of Banning Orders are punishable by imprisonment or fines depending upon the circumstances (see the Housing and Planning Act 2016).

In Chapter 3, the Housing and Planning Act 2016 makes provision for LAs to enter the offender's details on the Rogue Landlord database; this is a mandatory duty following imposition of a Banning Order, and there is no right of appeal. The procedure is discretionary where offenders have been convicted for a 'Banning Order offence' but a Banning Order was not sought or where they have committed two or more Banning Order offences during a 12-month period that were dealt with as Civil Financial Penalties instead of prosecutions. Within 6 months of conviction or issuing of penalties, a Notice of Intention must be served informing the offender that their details are to be entered. They may appeal to the tribunal within 21 days.

The database entry will include the offender's name and address, addresses of all properties owned, the offender's national insurance number and date of birth and details of the offences committed and details of the banning order if applicable. The LA are responsible for keeping the information updated. For banning orders, the entry remains as long as the ban is in force; for discretionary entries it is up to the LA how long the entry should remain.

Conclusion

Whatever direction a case may take, one thing remains constant: the evidence gathered and the statement written in support of it are crucial to its success. Cases will stand or fall depending on how investigators have collected, recorded and interpreted the evidence and how robustly the witness statement uses evidence to counter any reasonable excuse defence.

References

Andreou, C., Steinmann, S., Kolbeck, K., Rauh, J., Leicht, G., Moritz, S. and Mulert, C. (2018). The role of effective connectivity between the task-positive and task-negative network for evidence gathering [Evidence gathering and connectivity]. *NeuroImage*, 173, pp. 49–56.

Barratt, J. (2014). *Investigation and prosecution: Practical guidance for local authority enforcement officers*. London: Chartered Institute of Environmental Health.

Calvillo, D.P. (2014). Individual differences in susceptibility to misinformation effects and hindsight bias. *The Journal of General Psychology*, 141(4), pp. 393–407. https://doi.org/10.1080/00221309.2014.954917.

Contracts (Applicable Law) Act 1990. legislation.gov.uk Open Government Licence v3.0. Online. Available: www.legislation.gov.uk/ukpga/1990/36/contents (accessed 23 June 2021).

Criminal Justice Act 1967. s.9. legislation.gov.uk Open Government Licence v3.0. Online. Available: www.legislation.gov.uk/ukpga/1967/80/contents (accessed 26 June 2021).

Criminal Justice Act 2003 s.116. legislation.gov.uk Open Government Licence v3.0. Online. Available: www.legislation.gov.uk/ukpga/2003/44/section/114 (accessed 26 June 2021).

Gardiner, G. (2017). In defence of reasonable doubt. *Journal of Applied Philosophy*, 34(2), pp. 221–241. https://doi.org/10.1111/japp.12173

GOV.UK. (2021). Planning permission and building regulations. *Open Government Licence v3.0*. Online. Available: www.gov.uk/search-register-planning-decisions (accessed 23 June 2021).

Housing Act 2004 Part 1 s.30, s.32. Part 2. S.72 Part 3. s.95. Part 7. S.234, s.235, s.237, s.239(5)(7) and s.263. legislation.gov.uk Open Government Licence v3.0. Online. Available: www.legislation.gov.uk/ukpga/2004/34/section/237 (accessed 23 June 2021).

Housing and Planning Act 2016. legislation.gov.uk Open Government Licence v3.0. Online. Available: www.legislation.gov.uk/ukpga/2016/22/contents/enacted (accessed 26 June 2021).

Huq, S.F., Garety, P.A., and Hemsley, D.R. (1988). Probabilistic judgements in deluded and non-deluded subjects. *The Quarterly Journal of Experimental Psychology Section A: Human Experimental Psychology*, 40(4), pp. 801–812. https://doi.org/10.1080/14640748808402300.

Law of Property Act 1925 Part 1. legislation.gov.uk Open Government Licence v3.0. Online. Available: www.legislation.gov.uk/ukpga/Geo5/15-16/20/part/I (accessed 23 June 2021).

Local Government (Miscellaneous Provisions) Act 1976 s.16. legislation.gov.uk Open Government Licence v3.0. Online. Available: www.legislation.gov.uk/ukpga/1976/57 (accessed 24 June 2021).

London Borough of Haringey v Goremsandhu (2013). EWHC 3834 CO/4100/2013 Queens Bench Division. Lexis Library.

Ministry of Housing, Communities & Local Government. (April 2018). *Banning order offences under the Housing and Planning Act 2016 guidance for local housing authorities.* London: Open Government Licence.

Mishkin, P. and Moffatt, R. (2013). A review of multi-agency enforcement and discretionary property licensing to tackle Newham's private rented sector. In: Stewart, J. (ed.), *Effective strategies and interventions: Environmental health and the private housing sector* (pp. 12–14). London: Chartered Institute of Environmental Health.

Office of the Deputy Prime Minister. Housing Health and Safety Rating System Guidance (Version 2). ODPM Publications, West Yorkshire. (2006).

Police and Criminal Evidence Act 1984. legislation.gov.uk Open Government Licence v3.0. Online. Available: www.legislation.gov.uk/ukpga/1984/60/contents (accessed 26 June 2021).

Stewart, J. and Lynch, Z. (2018). *Environmental health and housing issues for public health* (2nd ed.). Abingdon: Routledge, Taylor and Francis Group.

12 Embedding research on public health and housing into practice

Matt Egan, Chiara Rinaldi, Jakob Petersen, Maureen Seguin, and Dalya Marks

Introduction

This chapter looks at the tricky question of how practitioners can use research on public health and housing to inform decisions affecting private rental housing. It will consider some of the evidence available and also discuss how practitioners can work with researchers to generate new evidence. The chapter will not pretend that evidence-informed practice is easy – in fact, we will go out of our way to discuss some of the challenges involved. It is our intention to present practitioners with (in our opinion) an honest summary of critical issues and highlight some practical ways forward.

We begin by considering why decisions about housing should be informed by health evidence. In the 1980s, the term 'evidence-based medicine' gained popularity, but examples of evidence being produced to inform decisions predate that decade – from both the health sciences and other areas of social policy and research (Petticrew, 2001). Ill health is strongly correlated with many forms of social disadvantage, including housing disadvantage (Marmot, 2020). Considering evidence of health impacts can help decision makers assess whether potential approaches to tackling housing problems are effective, ineffective, or harmful.

The policy analyst Carol Weiss (1979) has described this 'problem-solving' model of evidence use, noting that whilst researchers might like to think research utilisation should occur this way, what happens in practice is often quite different. For evidence informed decision-making to follow this problem-solving model, there needs to be evidence relevant to a particular decision, it needs to be reliable, and people need to be able and willing to use it. None of these preconditions can be guaranteed (Weiss, 1979).

For example, the UK's public health community spent much of the 1980s and '90s generating evidence to show that people with more money, higher social status, and other advantages were on average living longer and enjoying better health than people who were relatively worse off. Successive governments

DOI: 10.1201/9781003246534-13

treated this evidence with scepticism (Macintyre, 1997). When, at last, a government did accept that action needed to be taken to reduce these health inequalities, a new challenge presented itself. The public health community had focused on producing evidence that described the problem – but where was the evidence on how to solve it? Often, there was surprisingly little evidence to address the question *what works?* (Nutbeam, 2003; Petticrew et al., 2004).

Housing improvement and health – what is the evidence?

Research into housing and health exemplified this 'evidence gap'. In 2001, Thomson and colleagues (2001) published the first of their systematic reviews into the health effects of housing improvement. Systematic reviews are a useful way of understanding the evidence for a particular issue because these reviews aim to be comprehensive and emphasise the findings of more methodologically sound studies. Despite conducting an international search for housing improvement studies published over the previous 100+ years, Thompson and colleagues were surprised to find only eleven relevant studies – of variable quality – with inconsistent findings about health outcomes. At that time, this paucity of evidence seemed barely believable. After all, the links between housing and health had been a cornerstone public health for more than a century – to the extent that early UK government health ministries had often included housing within their remit. Surely, there *must* be evidence that housing improvement leads to health improvement!

There was. Just not that much. This mattered because housing improvement can be expensive and has the potential to lead to long-lasting impacts given the life span of the typical house or housing estate (Egan et al., 2013). The improvements themselves are varied. There are small interventions like new front doors or a paint job. Or larger improvements like new central heating systems, re-wiring, and retro-fitting. At their most radical, housing programmes can include demolition, relocation, new building, and redesigned neighbourhoods (Egan et al., 2015). Some housing interventions are universal whilst others target people with particular needs such as elderly or disabled residents, low-income households or homeless people. They might affect mental health, respiratory health, injury risk, other health outcomes, or other social outcomes that indirectly affect health (Thomson and Thomas, 2015). Individuals and communities might be harmed by poor-quality 'improvements', disruption, or relocation (Curtis et al., 2002; Egan et al., 2013). Or improvements may prompt gentrification – benefiting people who are already affluent, while disadvantaged residents are priced out of an improved area and obliged to move to poor housing elsewhere (Roshanak et al., 2018). All of these factors (and others besides) are things a decision maker working in housing may need to consider at some point. Eleven studies, with most of the best ones focusing on central heating improvements, barely scratched the surface of these many issues.

In 2013, Thomson and colleagues updated their evidence review (Thomson et al., 2013). In the intervening years, researchers in various parts of the world had been busy, and this time, thirty-nine studies were identified. Again, most of the best evidence identified focused on heating improvements – sometimes linked with complimentary measures such as ventilation. The reviewers concluded that these heating interventions were associated with improved respiratory health and other potential health benefits (e.g. mental health), especially where targeted at residents with inadequate warmth and those with chronic respiratory disease. For other types of housing intervention, there was some evidence of health benefits, but the quantity and quality of the evidence was poorer (compared to the heating studies), and findings were at times inconsistent or inconclusive. Furthermore, the review found little evidence on whether these interventions reduced health inequalities. Subsequent reviews by different authors have added to the sense that different kinds of housing improvement can, but do not always, lead to health improvement (McCartney et al., 2017). At the time of writing, the most recent large-scale review still concludes that there is relatively little evidence on whether housing improvements or other place-based interventions can reduce health inequalities (McGowan et al., 2021).

Housing improvement and health inequalities

The shortage of evidence on reducing health inequalities is a real problem. Since the creation of the welfare state, the health of the UK population has generally improved, but the relative gap between the better off and worse off has persisted and at times widened (Marmot et al., 2020). Part of the problem is that it is often easier to improve health outcomes for population groups who are already advantaged – they may be better able to respond positively to health messages and more able to benefit from interventions. We have already mentioned gentrification as one way in which this can happen for housing interventions. We need to know if those who benefit from a housing intervention are really the people with the greater need.

Once again, the problem is not an absolute absence of evidence. There is just not as much as we would like. A number of studies have found various initiatives to rehouse and support homeless people to be beneficial (Munthe-Kaas et al., 2018). Similarly, equipping houses to make them safer and more suitable for elderly or disabled people has at times been found to reduce injuries and other outcomes (CRD, 2014). Housing interventions targeting area-level disadvantage are less well evidenced, with some studies showing small impacts on inequalities. The GoWell programme, which looked at housing-led urban regeneration in Glasgow, produced what is currently the only study that assesses whether citywide investment in urban regeneration was allocated proportionally to need and whether this impacted on area-level health inequalities (Egan et al., 2015). It found that areas with poorer health did receive greater levels of investment in housing and neighbourhood improvement, and this did lead to relative reductions in area-level inequalities for mental and physical health outcomes. In principle, allocating investment according to need can be a way of improving health equity – even in cases (as with GoWell) where the implementers themselves were basing allocation decisions foremostly on housing needs rather than on health.

Housing policy and the private rented sector

In the world of housing, one particularly important dimension of inequality is housing sector: i.e. owner occupying, social rented, and private rented sectors. Each of these sectors includes examples of better- and poorer-quality housing, and a variety of people (better off and worse off) live in homes of each type. However, homes rented from private-sector landlords in the UK are more likely to be of poor quality compared to homes from the other sectors. Housing quality has improved across all sectors since 2000 but continues to be worse for private renters. The proportion of homes failing to meet the criteria of the Decent Homes Standard in 2019 was 23% in the private rented sector compared to just 12% in the social rented sector and 16% for owner-occupied homes (MHCLG, 2020). The UK's private rented sector accounted for 19% of all homes in 2019 – having doubled over the previous two decades in tandem with less affordable homes for sale and shrinking of the social housing sector. Hence, private rented homes have been both growing in number and tend to provide poorer living environments compared to homes from other sectors. This suggests a need for greater intervention and regulation of private rented homes to maintain and improve standards.

Once more, the research literature on the private rented sector and health is better at describing problems than evaluating solutions. Harris and McKee (2021) have reviewed this literature and found that substandard homes, with their risks of injury, other physical health problems (e.g. related to cold and damp environments), and mental health risks, are only part of the problem. Precarious housing, exacerbated by the ease of evicting tenants in the UK and poor affordability in some areas, also increases hardships for tenants and their families – directly affecting health and at times limiting opportunities that lead to indirect adverse health consequences (Clair et al., 2019).

The authors of this chapter conducted a literature search (led by Rinaldi) of evaluations that present findings on health outcomes of interventions affecting the private rented housing sector. We identified twenty-three studies. From a UK practitioners' perspective, these studies illustrate a common dilemma – the evidence most relevant to practice is not always the most methodologically robust (McGill et al., 2015). The most methodologically robust study we identified evaluated a US intervention in which vouchers were given to tenants to relocate to different areas with better-quality housing (Ludwig et al., 2012). Although this did lead to some health benefits, its applicability to a UK context is problematic, as this is not an intervention used here (currently). In contrast, the studies of UK interventions we identified were often small-scale evaluations with some useful qualitative findings on people's opinions about implementing particular initiatives – but less reliable quantitative evidence measuring impacts on specific outcomes. Around a third of the studies focused specifically on landlord licensing schemes in the UK (particularly 'selective licensing'; see Chapter 10). Some of the findings from these studies are also presented in Lawrence and Wilson (2019) *Independent Review of the Use and Effectiveness of Selective Licensing* and summarised in the panel that follows.

Panel 12.1: Summary of findings from Lawrence and Wilson (2019). *Independent Review of the Use and Effectiveness of Selective Licensing.*

Implementers should consider the following issues when planning and delivering selective licensing schemes.

- Realistic geographical boundaries that focus on problem areas
- Consider delivering as part of a wider suite of measures to support tenants
- Engagement with tenants and landlords
- Robust, consistent, and targeted inspection regime
- Collect and publish data on progress and outcomes

Challenges to implementation include:

- Difficulty identifying true extent of private rented sector in an area
- Non-compliance and rogue landlords
- Potential cost and administrative burden
- Legal requirement to renew after five years requires extra resource – administrative cost
- Secretary of state approval for large schemes can lead to extra delays
- The larger the scheme the more the potential cost and workload
- Inflexible schemes can lead to landlord resentment.

Embedding research into practice

So far we have summarised evidence relating to housing improvement and health. We have also made the point that the existing evidence has substantial limitations. Some issues have been researched more than others, and the research that has been done varies by quality and contextual relevance. Using evidence to inform decision-making therefore involves working with an imperfect evidence base. Systematic reviews are one useful tool for navigating this evidence and are regularly published. An internet search of your topic of interest along with the words 'systematic review' will often help you quickly identify a relevant review. The Cochrane Collaboration also has a repository of high-quality health-related systematic reviews (www.cochranelibrary.com), while the Campbell Collaboration performs a similar function for social science research (www.campbellcollaboration.org). The National Institute for Health and Care Excellence provides guidance that synthesises evidence reviews with expert consultations and has recently produced guidance on *Indoor air quality at home* (NICE, 2020).

Assuming that practitioners are able to identify evidence relevant to their decisions, there remains the question of how they will use this evidence. Practitioners have other competing sources of information – such as their personal experience and that of their colleagues as well as the public and other stakeholders (McGill et al., 2015). The range of decisions they can take may be limited by law, institutional structures, finance, and public opinion. Ideologies, beliefs, and values also shape decisions. The research community are unlikely to fully appreciate the context within which you work – and in some cases may have conducted their research in a very different context. Their recommendations can at times seem unfeasible – but remember that what is unfeasible now may not always be so.

Recognising some of these difficulties, there has been a shift in thinking from the 'problem-solving' model towards 'knowledge exchange'. The idea is that researchers and stakeholders (e.g. from policy, practice, and the public) should learn from each other (McGill et al., 2015). Researchers can benefit from this two-way exchange by better appreciating the decision-making context they seek to inform. If researchers and practitioners co-produce more relevant evidence at an appropriate time and in a format that suits decision makers, then this should help everyone.

Partnerships between research and practice can help practitioners evaluate their own projects and interventions. Sometimes, practitioners conduct their own in-house evaluations. Sometimes, external researchers provide advice – perhaps through membership of an advisory group. Some academics provide this function free of charge, though most are under pressure to bring in funding for the university, so their capacity for *pro bono* work is limited. Local authorities and other organisations can commission and pay for academics or private-sector research companies to conduct evaluations for them.

Evaluations can be quantitative (i.e. using statistics), qualitative (e.g. when participants discuss their views and experiences), or mixed methods (quantitative and qualitative). They can be designed to feed back information about how implementation is progressing to support continuous improvement (formative evaluations) or attempt more of a final verdict on how the intervention has performed (summative evaluations). They may seek to measure specific impacts (e.g. health, housing, or economic impacts) or examine barriers and facilitators to implementation (so-called 'process evaluations') or a combination of these (Skivington et al., 2021).

Evaluations require data. Often, relevant data is routinely produced and collected by implementing organisations, but practitioners should be aware that they may have to do some work themselves to identify and begin to prepare these data for the researchers. An alternative option is to collect bespoke data specifically for the evaluation, but this can drive up research costs and requires researcher involvement well before the intervention starts. Impact evaluations typically need pre-intervention ('before') data to form a baseline from which change resulting from the intervention ('after') is compared.

In order to separate intervention impacts from general trends, evaluators will also need similar data for a comparison group. Comparison groups should be as similar as possible to intervention groups but not exposed to the intervention being evaluated. In housing studies, this often means people living in similar homes and neighbourhoods but in different locations to the places where the intervention is rolled out. Good comparison groups are at times difficult to identify, and including them can drive up research costs. Still, they are usually essential for impact evaluations. Imagine trying to evaluate a three-year intervention that began in January 2019. Think about the huge effect that the COVID-19 pandemic has had on both housing and health since 2020 – it would be hard to even begin separating intervention impacts from COVID-19 impacts (or impacts of other factors such as new national policies, economic change, cultural change, etc.) without a good comparison group.

Many organisations do not have hundreds of thousands of pounds to spend on high-quality evaluations. When money is short, it may be possible to identify relevant freely available data or develop small evaluations that focus on specific issues that decision makers want to know more about.

For large-scale impact evaluations, it is possible to apply for funding – for example, the NHS National Institute for Health Research (www.nihr.ac.uk/) and the Economic and Social Research Council (www.esrc.ukri.org) fund many evaluations through various programmes. Funding applications can be lengthy processes, and the evaluations themselves can take two or three years to complete. Therefore, this should only be contemplated if the intervention has the potential to benefit large sections of the population (either directly or through replication on a wider scale) and a convincing research plan can be well specified in advance. Partnership with experienced researchers is essential for this.

Finally, remember that good evaluators publish *all* their findings – positive and negative. Sometimes, grand claims are made about housing interventions, but the housing system is complex and unpredictable. Be ready for bad news as well as good.

Conclusions

Evidence informed decision-making is generally considered good practice, but that does not mean it is easy. Identifying pre-existing evidence and conducting new evaluations are both challenging. Closer relationships between research and practice provide a sensible route for everyone involved to navigate this difficult terrain. For practice to be informed by research, we need research that is informed by practice.

The views expressed are those of the authors and not necessarily those of the NIHR or the Department of Health and Social Care.

References

Clair, A., Reeves, A., McKee, M., and Stuckler, D. (2019). Constructing a housing precariousness measure for Europe. *Journal of European Social Policy*, 29(1), pp. 13–28. https://doi.org/10.1177/0958928718768334

CRD (Centre for Reviews and Dissemination). (2014). Housing improvement and home safety. *Effectiveness Matters*, University of York. Online. Available: www.york.ac.uk/media/crd/effectiveness-matters-Nov-2014-housing.pdf (accessed 25 October 2021).

Curtis, S., Cave, B., and Coutts, A. (2002). Is urban regeneration good for health? Perceptions and theories of the health impacts of urban change. *Environment and Planning C: Government and Policy*, 20(4), pp. 517–534. https://doi.org/10.1068/c02r

Egan, M., KatikireddiI, S.V., Kearns, A., Tannahill, C., Kalacs, M., and Bond, L. (2013). Health effects of neighborhood demolition and housing improvement: A prospective controlled study of 2 natural experiments in urban renewal. *American Journal of Public Health*, 103, e47–53. https://doi.org/10.2105/AJPH.2013.301275.

Egan, M., Kearns, A., Curl, A., Katikareddi, S.V., Lawson, L., and Tannahill, C. (2015). Proportionate universalism in practice? A quasi-experimental study of a UK housing-led neighbourhood renewal programme's impact on health inequalities. *Journal of Epidemiology and Community Health*, 69, p. A49. https://doi.org/10.1136/jech-2015-206256.89

Harris, J. and McKee, K. (2021). *Health and wellbeing in the private rented sector Part 1: Literature review and policy analysis.* UK Collaborative Centre for Housing Evidence. Online. Available: https://housingevidence.ac.uk/wp-content/uploads/2021/07/HW-in-the-PRS-final.pdf (accessed 25 October 2021).

Lawrence, S. and Wilson, P. (2019). *An independent review of the use and effectiveness of selective licensing.* London: Ministry of Housing, Communities and Local Government. Online. Available: www.gov.uk/government/publications/selective-licensing-review (last accessed 25 October, 2021).

Ludwig, J., Duncan, G.J., Gennetian, L.A., Katz, L.F., Kessler, R.C., Kling, J.R., and Sanbonmatsu, L. (2012). Neighborhood effects on the long-term well-being of low-income adults. *Science,* 337(6101), pp. 1505–1510. https://doi.org/10.1126/science.1224648

Macintyre, S. (1997). The Black Report and beyond: What are the issues? *Social Science and Medicine,* 44(6), pp. 723–45. https://doi.org/10.1016/s0277-9536(96)00183-9

Marmot, M., Allen, J., Boyce, T., Goldblatt, P., and Morrison, J. (2020). *Health equity in England: The marmot review 10 years on.* London: Institute of Health Equity. Available: www.health.org.uk/publications/reports/the-marmot-review-10-years-on (last accessed 25 October 2021).

McCartney, G., Hearty, W., Taulbut, M., Mitchell, R., Dryden, R., and Collins, C. (2017). Regeneration and health: A structured, rapid literature review. *Public Health,* 148, pp. 69–87. https://doi.org/10.1016/j.puhe.2017.02.022

McGill, E., Egan, M., Petticrew, M., Mountford, L., Milton, S., Whitehead, M., and Lock, K. (2015). Trading quality for relevance: Non-health decision-makers' use of evidence on the social determinants of health. *BMJ Open,* 5, p. e007053. https://doi.org/10.1136/bmjopen-2014-007053

McGowan, V.J., Buckner, S., Mead, R., McGill, E., Ronzi, S., Beyer, F., and Bambra, C. (2021). Examining the effectiveness of place-based interventions to improve public health and reduce health inequalities: An umbrella review. *BMC Public Health,* 21, p. 1888. https://doi.org/10.1186/s12889-021-11852-z

MHCLG (Ministry of Housing, Communities and Local Government). (2020). *English Housing Survey: headline report 2019–2020.* London: MHCLG. Online. Available: https://assets.publishing.service.gov.uk/government/uploads/system/uploads/attachment_data/file/945013/2019-20_EHS_Headline_Report.pdf (accessed 30 March 2021; last accessed 25 October 2021).

Munthe-Kaas, H.M., Berg, R.C., and Blaasvaer, N. (2018). Effectiveness of interventions to reduce homelessness: A systematic review and meta-analysis. *Campbell Systematic Reviews,* 14(1), pp. 1–281. https://doi.org/10.4073/csr.2018.3

NICE (The National Institute for Health and Care Excellence). (2020). *Indoor air quality at home: NICE guideline [NG149]).* Online. Available: www.nice.org.uk/guidance/ng149 (accessed 25 October 2021).

Nutbeam, D. (2003). How does evidence influence public health policy? Tackling health inequalities in England. *Health Promotion Journal Australia,* 14(3), pp. 154–8. https://doi.org/10.1071/HE03154

Petticrew, M. (2001). Systematic reviews from astronomy to zoology: Myths and misconceptions. *British Medical Journal* (Clinical research ed.), 322(7278), pp. 98–101. https://doi.org/10.1136/bmj.322.7278.98.

Petticrew, M., Whitehead, M., Macintyre, S., Graham, H., and Egan, M. (2004). Evidence for public health policy on inequalities: 1: The reality according to policymakers. *Journal of Epidemiology & Community Health*, 58(10), pp. 811–816. http://dx.doi.org/10.1136/jech.2003.015289

Roshanak, M., Marra, G., Melis, G., and Gelormino, E. (2018). Urban renewal, gentrification and health equity: A realist perspective. *European Journal of Public Health*, 28(2), pp. 243–248. https://doi.org/10.1093/eurpub/ckx202

Skivington, K., Matthews, L., Simpson, S.A., Craig, P., Baird, J., Blazeby, J.M., Boyd, K.A., Craig, N., French, D.P., McIntosh, E., Petticrew, M., Rycroft-Malone, J., White, M., and Moore, L. (2021). A new framework for developing and evaluating complex interventions: Update of Medical Research Council guidance. *British Medical Journal*, 374, p. n2061. https://doi.org/10.1136/bmj.n2061

Thomson, H., Petticrew, M., and Morrison, D. (2001). Health effects of housing improvement: Systematic review of intervention studies. *British Medical Journal*, 323(7306), pp. 187–90. http://doi.org/10.1136/bmj.323.7306.187

Thomson, H. and Thomas, S. (2015). Developing empirically supported theories of change for housing investment and health. *Social Science and Medicine*, 124, pp. 205–14. https://doi.org/10.1016/j.socscimed.2014.11.043

Thomson, H., Thomas, S., Sellstrom, E., and Petticrew, M. (2013). Housing improvements for health and associated socio-economic outcomes. *Cochrane Database of Systematic Reviews*, 2. Art. No.: CD008657. https://doi.org/10.1002/14651858.CD008657.pub2.

Weiss, C.H. (1979). The many meanings of research utilization. *Public Administration Review*, 39(5), pp. 426–431. https://doi.org/10.2307/3109916

13 Developing effective PRS regulatory strategies

Russell Moffatt

Introduction

This chapter examines some of the important factors that lead practitioners might consider when developing and implementing effective PRS regulatory strategies.

A minority of landlords across the UK continue to commit housing crimes and expose tenants to life-threatening hazards and poor housing conditions. In response to this challenge, many local housing authorities are introducing regulatory schemes to increase enforcement against those that flout the law.

Based on observations of several successful regulatory schemes across England, a number of core areas have been identified as important strategic ingredients for success.

Clear vision

The first important step for a new scheme is establishing a clear vision for the intervention. Right from the start, policy makers and stakeholders should develop a clear narrative for schemes and stay focused on delivering the vision. There are many plausible justifications for LAs to develop a new regulatory strategy to tackle issues linked to the PRS, for example:

- Strong political mandate from local elected members based on their experiences within their constituency
- High levels of poverty linked to high concentrations of PRS
- Evidence of a serious PRS-related problem resulting in community stressors, including anti-social behaviour (ASB)
- Drive up standards where councils are discharging their homeless duty to the PRS
- Attempts to address housing market failure in very high- and low-demand areas

DOI: 10.1201/9781003246534-14

There are many possible justifications for new regulatory schemes. Whatever it might be, it must have a clear narrative that can be communicated widely and understood by a diverse audience. Most of all, it must be based on evidence.

Evidence

It is common for LAs to prepare an evidence base to support their proposals and present this to stakeholders. As part of this process, making a sound diagnosis of the housing issues, describing a plausible intervention, and setting objectives are important success components.

To assist with this process, it can be insightful to adopt a public health mindset, in this case poor housing conditions viewed as the disease. This process can be broken into two sub-stages.

Existence and distribution

The first step of designing a successful strategic intervention is understanding the distribution of poor housing and its symptoms in any given area. It is helpful to view this stage as a type of housing epidemiology. Establishing what and how many, where, and over what time an area is experiencing PRS stressors. These are simple questions that can often be frustratingly difficult to answer due to the hidden nature of many housing issues. Some of the most egregious internal housing conditions are masked by a neat exterior and a reluctance by landlords and tenants to allow the council access to private dwellings.

Historically housing authorities relied on street-by-street property surveys to develop an understanding of their PRS housing stock; however, this approach can be expensive and time-consuming, and the small sample size can skew results. As an alternative, there is now a range of data tools available to help authorities identify properties that are likely to be suffering from a range of housing stressors, including poor housing conditions and serious hazards (MHCLG, 2006). This process includes using pools of council-held and publicly available data to identify trends at the property level. Machine learning techniques have been developed to recognise data trends for each tenure type tailored for each council area. These data models can then be used to reliably predict tenure and other key stressors, including the location of housing hazards (Moffatt, 2020).

Property-level data can be overlaid with government produced data, normally available at a Lower Layer Super Output Area (LSOA) or ward level, to corroborate findings and help pinpoint areas that may require additional interventions.

A large-scale historical example of a detailed study of poverty and poor housing was Charles Booth's *Life and Labour of the People in London* (Booth, 1902). Booth's research helped reshape perspectives of poverty and poor housing through detailed property level research supported by case studies. Of course, it is not necessary to complete this level of research before commencing a new regulatory scheme. However, attempting to understand the reality faced by residents at the street and property level is important.

Understanding root cause

As with any disease, it is vital to understand what is causing it before an intervention is designed and made to treat it. To tackle the disease, we must first address the cause of the disease rather than treating the symptoms. Understanding the aetiology of the housing issues the scheme is seeking to address is likely to result in a more effective intervention. Insights provided by practitioners and case studies can reveal the common causal chains of certain poor housing outcomes and the impact on tenants' health and well-being.

An example of this approach in practice might be as follows. A landlord who fails to manage their property results in a rented property in poor condition. This is because dwellings degrade over time. Equipment and services fail (boiler), and materials wear out (floorboards split). These failures then result in serious hazards, for example excess cold and falls on a level (MHCLG, 2006). The longer the issues go unresolved, the worse the hazards get and the greater the exposure for the occupants, likely resulting in negative health outcomes. Using a causal chain analysis approach, it can be concluded that any negative health outcome is not root caused by the broken boiler or split floorboard. The root cause is further upstream, in this case the landlord who failed to maintain their property. By addressing the root cause, the physical hazards that are a vector for the poor outcome can be addressed in a more sustainable way.

Intervention design

We can use our understanding of the existence and distribution of poor housing and the causal chain to theoretically test against the widest range of possible enforcement intervention options. This might include reviewing best practice or developing a completely new intervention technique. It may also involve developing combinations of interventions to have the maximum impact.

Selecting the intervention scale can be challenging. Will it be across the whole borough or just in certain areas? Choosing where to intervene of course also means choosing when not to intervene, and this can be difficult for authorities to decide upon. Larger interventions of course require greater resources and political support.

Developing effective enforcement tactics will help the authority achieve meaningful enforcement outcomes. This includes identifying the right offence(s) to take enforcement action against and focusing enforcement resources to have the desired outcome. The development of a carefully thought-through enforcement policy will aid this process.

Most tenants suffering criminal housing conditions and management will tend not to complain to housing authorities (Citizens Advice, 2018). Therefore, proactive work to identify the locations of housing crimes is essential. A reactive approach requires fewer resources; however, it is unlikely to achieve any meaningful behaviour change amongst the landlord population.

Studying local non-compliant landlord behaviour can assist with intervention design. For example, landlords responsible for criminal housing tend not to license their properties. Therefore, find the unlicensed PRS properties, and it is likely that tenants suffering criminal housing will also be found (Moffatt, 2018).

Also, consider the use of behavioural wedges to help separate landlords that just need a nudge from those that have entrenched non-compliant attitudes. This enables regulatory teams to deploy their limited resources where they can be most effective. An example of this approach is to strategically use requests for safety certificates as a way of testing compliance and to determine if further regulatory action is required to achieve the behaviour-change objective.

Maintaining a focus on tenants' welfare during the enforcement process will help to protect against negative impacts, particularly when vulnerable tenants are involved. Incorporating the views of private tenants helps improve the enforcement outcome. However, these also need to be balanced against the wider benefits for other residents.

Finally, any new intervention should be piloted to observe how it works in the field and its effectiveness measured. It also provides intervention teams with an opportunity to hone tactics and mitigate any negative impacts.

Monitoring and communication

Once the intervention design is decided upon, the final step is to assess how the impact of the regulatory interventions can be measured and assessed over the longer term. Ideally any measure of an intervention should be focussed on outcomes. However, this can be problematic, because many powerful confounding factors often intersect with the proposed intervention outcome, including health, wellness, income, crime, etc. Therefore, a blend of outputs and outcomes based on a reasonable baseline is normally a good compromise. Also, it is important not to submit to objective overload; avoid using one intervention to solve all the issues a local area faces.

Once a regulatory scheme is underway a lead practitioner should keep developing the narrative and maintaining the stakeholder consensus. This includes sharing scheme successes with the wider public and maintaining a dialogue within and outside of the housing authority. Scheme media can include monthly performance updates, newsletters, articles, and TV and radio programmes.

For regulation to be effective those undetected individuals who might be minded to commit housing offences need to learn that there may be a serious penalty for poor behaviour. The importance of communicating this message through media channels should not be underestimated.

Intervention delivery

Organisational resources have a significant impact on the size and nature of their interventions. The very best intervention design can be rendered ineffective by poor delivery due to a lack of leadership, skills, or resources.

Many regulatory teams have a core of officers from which larger teams can be developed. Matching and balancing practitioner skills and experience helps produce a high-performing team. Other factors such as motivation, diversity, and adaptability all play an important part and deserve serious consideration.

Setting up and delivering a new regulatory scheme requires leadership and careful planning by the lead practitioner. A successful scheme setup necessitates multidisciplinary project management, including legal, policy, recruitment, data, technology, and service redesign, often all at once. This can place significant strain on services if not well planned (Moffatt, 2019).

Having a sound understanding of how LAs really work can be helpful, including how to navigate the decision-making channels, understanding where power rests and how to unlock resources. This includes legal and finance departments whose job in part is to ensure front-line teams stay within the law and budgets. These internal pressures, if not carefully, managed can negatively impact on schemes and their ability to achieve objectives. Therefore, learning to navigate and negotiate with techno-structures (Mintzberg, 1992) can be important to a scheme's overall performance.

Role of technology

Effective modern enforcement often depends on embracing new technology. It can drastically improve productivity and provide insights not accessible by even the most accomplished practitioners. Adopting the right technology should be done as part of the scheme setup and can be a major success factor if well managed.

As one example of how enforcement can be boosted by technology, practitioners at Newham Council used big data and machine learning techniques to identify 10,000 unlicensed and unsafe private rented properties. It allowed the enforcement team to be much more productive by minimising wasted time by focusing on addresses where there was a much higher likelihood of poor housing conditions. In just a couple of years Newnham's private housing enforcement outputs went from 25 prosecutions to 250+ prosecutions per year, a 10-fold increase. In 2016 this represented 67% of all private housing enforcement undertaken in London (Moffatt, 2020)

Multiagency partners

It is not unusual to find that criminal landlords are breaking the law across the piece, including planning, building control, tax, fraud, court orders, etc. Moreover, criminal housing can also be a home for individuals who are of interest to the authorities. This makes PRS enforcement ideal territory for a multi-agency approach. Enforcers are more effective when they work together. Practitioners should attempt to deal with the cause of poor housing and the symptoms of poor housing, including ASB and other issues within a single visit, if possible. This can mean working at a granular level with police, Immigration Enforcement, ASB Service, Fire Authority, HMRC, Homelessness teams and charities, Planning, Building Control, Gangmasters, and Labour Abuse Authority, etc. (Moffatt, 2019).

Greater intelligence sharing within the council and across agencies is crucial. Intelligence-driven enforcement allows multi-agency teams to focus on properties which link to other crimes. Adopting the principles of disruptive policing (Kirby et al., 2015) is gaining traction amongst LAs eager to explore how they can better manage the PRS through focused use of intelligence and regulation (Stewart and Moffatt, 2019)

This approach can magnify the enforcement impact. For example, Housing Act prosecutions and financial penalties, in conjunction with Planning enforcement and Council Tax Court Orders, can have an overwhelming legal impact on a determined criminal landlord.

Successful multi-agency working is dependent on developing many productive working relationships with the widest range of agencies. Identifying officers who make professional relationships quickly and effectively can help speed this process up. Multi-agency cooperation works best when organised at the field level with strategic level support (Moffatt, 2019).

Conclusions

There are several complex factors to be considered when developing effective PRS regulatory strategies. Many of the steps can be challenging for lead practitioners to overcome. However, it is possible to navigate these challenges by adopting a public health approach, embracing technology, and working within a wider regulatory team.

References

Booth, C. (1902). *Life and Labour of the People in London*, *LSE*. Online. Available: https://booth.lse.ac.uk/learn-more/what-was-the-inquiry (accessed October 2021).

Citizens Advice. (2018). Touch and go, How to protect private renters from retaliatory eviction in England. *Citizen Advice*. Online. Available: www.citizensadvice.org.uk/Global/CitizensAdvice/Housing%20Publications/Touch%20and%20go%20-%20Citizens%20Advice.pdf (accessed October 2021).

Kirby, S., Northey, H., and Snow, N. (2015, December). New crimes – new tactics: The emergence and effectiveness of disruption in tackling serious organised crime. *The Journal of Political Criminology*, 1(1), pp. 33–44.

Ministry of Housing, Communities and Local Government. (2006). *Housing health and safety rating system (HHSRS) operating guidance: Housing inspections and assessment of hazards*. London: MHCLG. Online. Available: www.gov.uk/government/publications/hhsrs-operating-guidance-housing-act-2004-guidance-about-inspections-and-assessment-of-hazards-given-under-section-9 (accessed October 2021).

Mintzberg, H. (1992). *Structure in fives: Designing effective organizations*. New Jersey, US: Pearson Education.

Moffatt, R. (2018). Does property licensing improve property standards? *Metastreet*. Online. Available: https://metastreet.co.uk/blog/does-property-licensing-improve-property-standards.html (accessed October 2021).

Moffatt, R. (2019). The building blocks of effective private housing multi-agency enforcement. *Metastreet*. Online. Available: https://metastreet.co.uk/blog/the-building-blocks-of-effective-private-housing-multi-agency-enforcement.html (accessed October 2021).

Moffatt, R. (2020). Predicting the location of housing crimes – The power of machine learning and property licensing. *Metastreet*. Online. Available: https://metastreet.co.uk/blog/predicting-the-location-of-housing-crimes.html (accessed October 2021).

Stewart, J. and Moffatt, R. (2019). *Rooting out the rogues: 'targeted disruption' as a new approach to tackling issues in England's growing private rented housing sector?* Online. Available: https://metastreet.co.uk/files/Rooting.out.the.rogues_targeted.disruption.pdf (accessed October 2021).

Conclusions

This book demonstrates how complex the private rented sector is and how challenging it can be for practitioners and policy makers to intervene effectively. The PRS market is diverse, constantly in flux and subject to a piecemeal regulatory framework with little overall policy coordination. It remains the case that a minority of the PRS tenants are exploited by criminal landlords, adversely affecting their health, safety and security.

Many local authorities have responded with radical interventions, including PRS licensing, in part because it offers self-funded mechanisms for improving property standards and excluding the worst offenders. Setting up local PRS licensing schemes is a resource-intensive activity that can lead to increases in enforcement and regulation. However, large-scale licensing remains contentious, and local authorities continue to encounter opposition mainly from landlords.

It is crucial that effective research and evaluation expand to help capture PRS intervention successes and failures. This will support and help direct future strategic interventions. Encouraging practitioners to engage with academic research and adopt public health approaches will improve effective intervention design and generate new evidence.

EHPs operate within a complex framework and require multiple skills and advanced knowledge to promote and enforce better conditions in the PRS. This often includes protecting security of tenure for tenants. EHPs have a wide range of opportunities to work in partnership with public, private and third-sector organisations. A joined-up approach enhances interventions and helps deliver meaningful improvements for private tenants and the wider community. Developing intervention evidence will help professionals navigate the maze of legal, operational, strategic, policy and technology challenges they currently face.

Many authorities face high-risk scenarios related to HMOs, including fire safety, serious disrepair and over-crowding. These can be very challenging to regulate effectively. Each LA will have its own PRS strategy utilising criminal

DOI: 10.1201/9781003246534-15

and civil remedies. A range of options is available, and the practical and effective application of each intervention need to be carefully considered. Working in partnership with other organisations as part of an overarching enforcement strategy can help deliver meaningful change for tenants and the wider community.

As the PRS market evolves, so do the tools practitioners have to address poor housing, including HHSRS, civil penalty notices, banning orders, Rent Repayment Orders and Management Orders. Laws and regulations are dynamic and require continued investment in staff training to help maintain workforce competency. Complex enforcement interventions stand or fall depending on how well practitioners collect, record and interpret gathered evidence. Complex court structures and the duty for practitioners to act in a clear and transparent manner guided by LAs' policies add to the challenge.

It is our hope that this book will provide both inspiration and impetus to the workforce and other stakeholders charged with developing and delivering evidence-based approaches to better regulating the privately rented housing sector.

Index

banning orders 2, 27, 45–46, 59, 62, 97, 102–103, 124

civil remedy and penalties 2, 24, 27, 40–45, 58–62, 65, 79, 89, 101, 103, 124
coastal towns, seaside towns 9, 17–18
cold and damp (fuel poverty, energy efficiency) 34–36, 69–73, 80, 82, 109, 118
courts and tribunals 26, 40–47, 53, 55, 60–61, 68, 70, 77, 81, 86, 89–91, 99–100, 102–103, 121, 124
criminal landlord (rogue landlord) and remedy 24–25, 40–42, 48–55, 58–60, 62, 65, 79, 97, 101–102

demand for private rented housing 7–8
deposit protection 52

Environmental Health Practitioner (EHP) 2, 14, 22–26, 30–31, 33–38, 66, 68–73, 123
eviction 16, 26, 27, 48–49, 52–55, 62, 109
evidence gathering 97–105, 117–118

fuel poverty (cold and damp, energy efficiency) 34–36, 69–73, 80, 82, 109, 118

geography and private rented housing 8–10, 17–19

harassment and illegal eviction 53–54
hoarding 37
home, meaning of 14–21
hospital discharge 35

house in multiple occupation (HMO) 1, 7, 9, 16, 26–27, 35, 42–44, 54–55, 59, 76–87, 88–89, 92, 100–101, 103, 123
Housing Act 2004 2, 11, 26, 42–44, 60, 62, 63, 67, 76, 78, 80, 82, 88–94, 98–101, 226
Housing and Planning Act 2016 2, 11, 26–27, 58, 62, 63, 102, 103
Housing Health and Safety Rating System (HHSRS) 66–75

interventions 1–2, 10–12, 17–18, 22–25, 27, 30–38, 58–65, 80, 89, 107–113, 116–122
investigation 43, 51–52

landlords and agents 1–2, 5–6, 8–11, 16, 19, 25–28, 35, 40–41, 43–46, 48–56, 58–65, 66, 68–70, 78, 80, 82, 88–89, 91–95, 97–98, 101–103, 109–110, 116–119, 121, 123
licencing 2, 26–28, 35, 43–44, 58, 61–64, 78–81, 84, 86, 88–96, 99, 101, 103
life-course 17–19, 31

management and manager 1–2, 5–6, 9, 11, 14, 19, 24–28, 51–53, 61, 63–65, 76, 78–81, 83, 86, 88–89, 92–93, 98–101, 103, 118–121, 124
Management Orders, Interim and Final 58–59, 63–65

offences 43–45

partnerships 30–39
planning and HMOs 81–82

powers of entry 98–100
private rented sector 1–3, 4–13,
    48–49
property licencing 88–96
prosecution 45–46
public health and housing 30–39,
    106–115

registration 28
regulatory framework 10–12, 22–29
regulatory skills 97–105
Rent Repayment Order 2, 27, 52–53,
    58, 62–63, 124
rent-to-rent 53–54
research into practice 106–115
Rogue Landlord database 102–103

seaside towns, coastal towns 9, 17–18
security 1, 14–17, 19, 22, 25, 28, 48,
    88, 123
shadow private rented sector 48–57
strategies 1–2, 18, 24, 30, 33, 58, 60,
    95, 116–122
supply of private rented housing 5–6

tenant and tenancy 1–2, 6, 10–12,
    14–19, 22, 25–28, 35, 40–41, 44,
    48–53, 55–56, 59. 62–66, 66, 69–71,
    78, 82, 88, 94–95, 97–103, 109–110,
    116–119, 123–124
tribunals and courts 26, 40–47, 53, 55,
    60–61, 68, 70, 77, 81, 86, 89–91,
    99–100, 102–103, 121, 124

CPSIA information can be obtained
at www.ICGtesting.com
Printed in the USA
BVHW050021100522
636354BV00002B/28

9 781032 159690